# The Joy of Insight

THIS BOOK IS PUBLISHED AS PART OF AN ALFRED P. SLOAN FOUNDATION PROGRAM

# THE JOY
# OF INSIGHT

❖

## Passions
## of a Physicist

❖

VICTOR WEISSKOPF

BasicBooks
*A Division of* HarperCollins*Publishers*

Material from letters between Kapitza and Dirac copyright © 1990 by the USSR Academy of Sciences and reprinted with the permission of General Media International, Inc.

Material from *The Advisors* by Herbert York reprinted by permission of W. H. Freeman and Co., New York.

Translation by Victor Weisskopf of lines from *The Threepenny Opera* based on Grove Weidenfeld U.S. edition published 1964, by permission of the publisher.

Material from the scientific correspondence of W. Pauli reprinted with the permission of Springer-Verlag, Heidelberg.

Originaltitel "gott spricht," aus: Wilhelm Willms, der geerdete himmel © Verlag Butzon & Bercker, Kevelaer/West Germany, 7. Aufl. 1986.

Photograph of proton synchrotron reprinted by permission of the Conseil Européen pour la Recherche Nucléaire, and the AIP Niels Bohr Library. Photos of Erwin Schrödinger and of Eugene Wigner and Henry Barschall reprinted by permission of the AIP Niels Bohr Library.

Library of Congress Cataloging-in-Publication Data
Weisskopf, Victor Frederick, 1908–
    The joy of insight : passions of a physicist / Victor Weiss-
kopf.
      p.    cm. — (Alfred P. Sloan Foundation series)
    Includes index.
    ISBN 0-465-03678-3
    1. Weisskopf, Victor Frederick, 1908–   .  2. Physi-
cists—United States—Biography.   I. Title.   II. Series.
    QC16.W516A3   1991
    530'.092—dc20
    [B]                                  90-49443
                                         CIP

*To my dear wife, Ellen,*
*with whom I shared my life from 1932*
*until her death on 6 August 1989.*

The joy of insight is a sense of involvement and awe, the elated state of mind that you achieve when you have grasped some essential point; it is akin to what you feel on top of a mountain after a hard climb or when you hear a great work of music.

—Victor Weisskopf

# Contents

## Contents

# Preface to the Series

THE ALFRED P. SLOAN FOUNDATION has for many years had an interest in encouraging public understanding of science. Science in this century has become a complex endeavor. Scientific statements may reflect many centuries of experimentation and theory, and are likely to be expressed in the language of advanced mathematics or in highly technical terms. As scientific knowledge expands, the goal of general public understanding of science becomes increasingly difficult to reach.

Yet an understanding of the scientific enterprise, as distinct from data, concepts, and theories, is certainly within the grasp of us all. It is an enterprise conducted by men and women who are stimulated by hopes and purposes that are universal, rewarded by occasional successes, and distressed by setbacks. Science is an enterprise with its own rules and customs, but an understanding of that enterprise is accessible, for it is quintessentially human. And an understanding of the enterprise inevitably brings with it insights into the nature of its products.

The Sloan Foundation expresses great appreciation to the advisory committee. Present members include the chairman, Simon Michael Bessie, Co-Publisher, Cornelia and Michael Bessie Books; Howard Hiatt, Professor, School of Medicine, Harvard University; Eric R. Kandel, University Professor, Columbia University College of Physicians and Surgeons, and Senior Investigator, Howard Hughes Medical Institute; Daniel Kevles, Professor of History, California Institute of Technology; Robert Merton, University Professor Emeritus, Columbia University; Paul Samuelson, Institute Professor of Economics, Massachusetts Institute of Technology; Robert Sinsheimer, Chancellor Emeritus, University of California, Santa Cruz; Steven Weinberg, Professor of Physics, University of Texas at Austin; and Stephen White, former Vice-President of the Alfred P. Sloan Foundation. Previous members of the committee were Daniel McFadden, Professor of Economics, and Philip Morrison, Professor of Physics, both of the Massachusetts Institute of Technology; George Miller, Professor Emeritus of Psychology, Princeton University; Mark Kac (deceased), formerly Professor of Mathematics, University of Southern California; and Frederick E. Terman (deceased), formerly Provost Emeritus, Stanford University. The Sloan Foundation has been represented by Arthur L. Singer, Jr., Stephen White, Eric Wanner, and Sandra Panem. The first publisher of the program, Harper & Row, was represented by Edward L. Burlingame and Sallie Coolidge. This volume is the ninth to be published by Basic Books, represented by Martin Kessler and Susan Rabiner.

—The Alfred P. Sloan Foundation

# Preface

I WAS BORN and brought up in central Europe and was transported to America by world events. I have lived through two world wars—one as a child, the other as a bomb maker—and ended up as a scientist respected on both sides of the Atlantic. In spite of all the wars and the political strife that surrounded me, I have lived a charmed life. My immediate family and I never suffered any harm. We were saved by good luck and the help and support of the international family of scientists.

In my youth I had the opportunity to absorb the best of the European culture in science, art, and music, and although I witnessed the destruction of that culture by the Nazis, I also saw its resurrection after the war.

Despite its political and cultural earthquakes, this century has been one of unprecedented development in scientific knowledge. Indeed, for a scientist caught up in the events of the last sixty years Charles Dickens's words ring true: "It was the best of times, it was the worst of times."

I have been privileged to have gained many insights into scientific matters and also into other forms of creative expression in culture and human affairs. I have written this book to describe what I have experienced and especially to share with the reader the pleasures I have had in my work and my life, as well as my concerns during the many periods of trouble during which I have lived.

The autobiography of a scientist must contain a good deal of science. I have endeavored to keep those parts simple enough for readers who have only a rudimentary acquaintance with science. I may not have succeeded, but I tried very hard to keep it understandable without sacrificing too much of the excitement I felt at what I have called "the joy of insight."

I have relied mostly on my memory while writing about my life and my ideas. I have never had a systematic file of letters or other documents, nor have I kept diaries. My numerous changes of residence, together with my aversion toward neat filing or registering of events, have made it hard to check the accuracy of certain happenings, but I am reasonably sure that I have been truthful in describing the events in this book and the role of the people involved. Any errors I have made are due solely to a lapse in my imperfect memory.

I should also say that when I describe events, people, or ideas that I have already written about in previous essays, I have sometimes plagiarized myself to some extent because I could not think of a better way of expressing myself. In particular, this is true in the case of the characterizations of Wolfgang Pauli in chapter 5 and of Werner Heisenberg in chapter 8 and also in the last chapter when I formulate my approach to general problems of human concern. Some of the phrases used in these parts are almost identical to those in two previous essays that appeared in a collection of my articles entitled *The Privilege of Being a Physicist,* published by W. H. Freeman and Company in 1989.

I am very grateful to the Alfred P. Sloan Foundation for having suggested this autobiography and for their encouragement and support during the writing. I am especially indebted to Lucie Prinz for her indefatigable help and her valuable advice regarding content and style and for the actual production of the manuscript. I would not have been able to write this book without her collaboration. I also would like to express my thanks to my son-in-law, Douglas Worth, for his help in the translations of German passages into English.

# ONE

⊠

# Formative Years in Vienna

On 10 AUGUST 1923 I spent a night on the Loser, a mountain in the Austrian Alps. My friend George Winter was with me. In our backpacks we carried enough food to get us through the night, but our real hunger was for scientific information. August 10 is the night on which the annual Perseid showers of shooting stars are at their height. From the vantage point of our mountain perch George and I planned to study the stars as they fell. He was seventeen years old. I was fifteen.

We sat back to back, our eyes focused on the heavens. As the stars shot through the skies, we wrote down their color (red, yellow, white) and noted whether the light they gave off was weak or strong. Throughout the night we filled our notebooks with figures, our hands getting stiffer and colder by the hour. But when we climbed down at dawn, exhausted, we had the raw data for a scientific paper. We sent our findings to an organization of amateur astronomers, Der Bund der Sternfreunde (The Friends of the Stars).

They rejected our paper. It was much too important for their small amateur newsletter, they said. With our permission, they would send it off to the *Astronomische Nachrichten* (*Astronomical News*), a serious scientific publication for professional astronomers. We were astounded, and we were even more surprised when it was accepted and appeared in print. Weisskopf and Winter had published their first scientific work. We felt that we were launched into our professional lives. (George became an engineer.) To this day our paper on the shooting stars remains at the top of our publication lists.

I had become obsessed by astronomy during that year, and my father had given me a telescope. It was small but good—by no means a toy. For some time I had been subscribing to *Kosmos,* a general scientific magazine that contained popularized articles about diverse subjects. I also read the books *Kosmos* published, and I had taken courses in science at Urania, an adult education center. Our teacher, Mr. Thomas, took the class into the outskirts of Vienna at night to look at stars. I was captivated and memorized as many of the constellations as I could. (Even today I can beat many astronomers with what I learned about the constellations as a teenager looking up at the stars that shone down on Vienna.)

However, my interest in science in general had begun when I was even younger. Before I had graduated from the gymnasium, I knew that I wanted a life in science. When I was a bit older, I devoured everything that had to do with relativity and atomic structure, two fields that were in the process of development at that time. I understood only part of what I read, but I was fascinated by the new ideas.

My parents supported me by buying the books I wanted, but they could not help me intellectually. The science teachers at my gymnasium were not much more help, since they did not yet understand these subjects. I was lucky enough to have a few friends, a little older than I, with whom I could discuss some of these problems. Among these was Otto Frisch, who later played an important role in the discovery of nuclear fission.

One question defeated me, and I could find no one to help me with the answer. It had to do with some consequences of relativity and quantum theory. Einstein's relativity theory teaches us that an object moving at the speed of light should have an infinite energy. This is why ordinary particles can approach but never reach the speed of light. But what about the quanta of light? Einstein found in 1905 that there is strong evidence in favor of considering a light beam not as a continuous stream of energy but as a bundle of particles, which he called light quanta. These quanta move with light velocity, but we know that their energy is not infinite. It is finite and rather small and depends on the color of the light.

When these questions obsessed me, I was only fourteen years old. In the summer of 1922, with the chutzpah of the very young, I decided to get the answer to my question from the man who founded quantum theory, the man whose name is forever associated with quanta. I wrote a letter to the great Max Planck telling him that I hoped he could help me understand the problem.

Imagine my joy and excitement when, a few weeks later, I received a postcard from Planck in which he tried to explain the discrepancy I had been struggling with. The facsimile of that card is found in the photo insert. Here is the English translation:

Dear Sir! I will try to answer your question, expressed in your letter of the 25th of last month as I understand it. The statement that a mass [Planck uses the term *mass* as a measure of energy] moving with light velocity becomes infinitely large is valid only for corpuscular (convective) motions, such as matter or electrons, but not for wave motions; one must distinguish in principle between the speed of convection and the speed of propagation. The motion of light quanta has to be considered as a wave motion. Therefore one must not conclude that their masses are infinitely large.

Respectfully yours,
Dr. Max Planck

The style of the note ("Dear Sir!" and "Respectfully yours") shows that I had succeeded in hiding my age when I wrote to him. This document is not just a personal memento. It has a special historic significance for a deeper, more universal reason: Planck's answer was wrong! Not that I was able to recognize this when I received his message. All I could think of was that I was being taken seriously by one of the greatest physicists of the time. His response encouraged me more than anything to become a physicist. But the fact remains that Max Planck's answer to my question is not correct.

I could not have known that my question was very much on the minds of the best scientists of that time. Quantum theory was in a crisis. The particle nature of light was not generally accepted and understood. A discovery in 1922 changed the whole situation: the American physicist Arthur Holly Compton demonstrated the particle nature of light by analyzing the collisions of light with electrons. Two years later the French physicist Louis de Broglie suggested that ordinary particles such as electrons also have wave properties, a fact that soon after was shown to be correct and made Planck's answer to me invalid. The correct answer would have been difficult to arrive at before Compton published his findings. It should have been this: Only particles that have a rest mass (i.e., a measurable mass when they are at rest), acquire infinite energy when moved with light velocity. The light quantum has no rest mass: indeed, light can never be put to rest. It always moves with light velocity. Had I written my letter a year later, I would have received a more accurate answer.

Planck's postcard came to me in the little resort village of Altaussee, not far from Salzburg, where my family spent every summer in a house called Villa Charlotte. It was named for our family's leading figure, my great-grandmother Charlotte Cohn, to whom it had belonged. "Die Urmama," as she was always called, was a formidable woman. She died when I was three years old, but she belongs in the story of my life because her influence on my family was profound. In fact, her spirit was such a presence in our lives that I have always felt as if I had actually known her well when I was a child.

Charlotte Cohn was born in 1828, a year after Beethoven's death, at a time when women were not expected to exert much power except over their small domestic domain, their children and their household staff. If a woman became wealthy, it was usually because she had received a large dowry from her parents or because she had married a rich man. Die Urmama, raised in a moderately well-to-do family, became rich, but she accomplished this feat by being a shrewd and successful businesswoman. At a relatively young age, she took over a failing bank and turned it into a financial success. In real estate transactions she was so astute that when she died she owned a great deal of valuable property, including the Altaussee summer house and the large house in Vienna where I grew up.

Probably quite deliberately, she married a mild-mannered, pleasant man who did not have a particularly forceful personality and willingly let her run the show. They had two daughters and a son. One of the daughters, who was often described as disturbed, traveled widely and led a rather aimless life. She never married. By contrast, her sister, my grandmother, was a proper Viennese girl who did what was expected of her. She married a Hungarian businessman named Alexander Gut. My memories of him extend not much further

than the nice presents he always gave us at Christmas. Martha, my
mother, was my grandparents' only child. They provided her with
the easy, comfortable life of a Victorian girl of the upper middle
class. She was sheltered and protected with the usual nannies and
servants to smooth the way for her.

My father, Emil, came from a different background. He was born
in Sucice, a small town in Czechoslovakia. At that time it was still
part of the Austro-Hungarian Empire and had the German name of
Schüttenhofen. Within its walls there was a Jewish ghetto where
nearly a quarter of the town's population lived. Emil's father was
the kosher butcher, the *schochet*. As a small boy, Emil got up before
dawn to hold a candle for his father while he severed the animals'
arteries as required by Jewish law.

He was the only boy in a family of several girls. In the local school
he was the prize student. The rabbi, recognizing that a boy with his
intellectual gifts would be limited by the education Sucice offered,
took up a collection to send Emil to Pilsen to continue his studies at
the gymnasium. Emil did not disappoint the rabbi. In his eight years
at the gymnasium in Pilsen he had a perfect record. His academic
standing should have earned him a full scholarship to the university,
a reward traditionally presented to the top students by the emperor,
who also awarded each of the prize students a golden ring. But the
responsible officials did not want to accord such honors to a Jew
from a shtetl, and Emil was cheated out of his reward.

Undaunted, he went to Vienna, determined to work his way
through the university by tutoring the sons and daughters of the
wealthy. After Emil received his doctor of law degree, he passed the
necessary examinations and became a district judge in a suburb of
Vienna, the lowest step in his chosen career as a jurist.

One evening he was invited to a party at my mother's family's
apartment. There he met my mother, and they soon fell in love. Die
Urmama was not pleased. Her assimilated, nonobservant Jewish
family prided itself on its enlightened outlook. They were Austrians
who happened to be Jewish. That Emil was a district judge was not
as important as his background. The fact remained that he was a
poor young man from a tiny shtetl. His father was a butcher—and
a kosher butcher at that. It wouldn't do at all.

A familiar story unfolded. The two lovers met secretly. They wrote letters of love and longing. There was even talk of an elopement, but in 1904, when Emil was thirty-one and Martha was twenty-four, they married in a ceremony attended by the family. (From what I know of her, I doubt that Die Urmama attended.) At first the newlyweds were not invited to set up their household in Die Urmama's house, as would have been natural in our family. It was several years before she finally relented. Then they were admitted to the family mansion. Indeed, my father became her trusted legal advisor during the last years of her life.

After the wedding my father decided to switch careers and become a lawyer. He had to start at the bottom again, with a clerkship at a highly respected law firm. In time he became a partner in the firm and during his life he was known as a most successful attorney. Exactly nine months after their marriage Martha gave birth to their first son, my brother, Walter. I followed four years later, and my sister, Edith, was born two years after that.

At that time, Vienna was still the capital of the Austro-Hungarian Empire, whose citizens were people of many different nationalities and cultures ruled by the monarch Kaiser Franz-Joseph. Six years after my birth, the First World War broke out. The monarchy, allied with Germany, lost the war. The Austro-Hungarian Empire broke up into small national entities. The German-speaking part became the Republic of Austria, with Vienna as its somewhat overgrown capital.

My family lived in a large apartment of my great-grandmother's house, situated in a well-to-do neighborhood near the city hall and the charming Rathaus Park, a lovely place with well-kept gardens, playgrounds, flower beds, and trees. Thanks to Die Urmama's resources and my father's successful law career, we were reasonably well off, with three servants, a cook, a chambermaid, and a nurse for the children. During my early childhood we spent a lot of time in Rathaus Park. We were well supervised by the nurse, who warned us not to play with "bad children"—meaning those of the lower classes, the sons and daughters of the servants in the big houses. In later years we had fewer servants. The cook was the first to go. His duties were taken over by Mother and the maid. Later the nurse left, and there was only one person to help run the household and watch the children. In general, however, those were very protected times for the children of well-to-do families. The problems of the day were kept far from us.

Because my father was so often more of an absence than a presence in my life, I don't have as many memories of him as I have of my mother, but one incident stands out. I was very small, perhaps around five, when he decided to take his family to the shtetl where he was born to show us to his relatives. We traveled to Sucice in a hired car with a chauffeur. A car was still quite a novelty at that time, and along the way there were many mishaps. At one point water was put into the gas tank by mistake, delaying us for a bit.

Finally we arrived in Sucice. We must have been quite a sight as we pulled up to my father's family home. I remember most vividly a great many women who hugged and kissed us. Some of them had hair on their faces. They were my father's sisters and aunts, and I can still remember the strong smell of garlic and other unfamiliar foods that enveloped us as they embraced and kissed us. My elegant, somewhat reserved mother must have felt completely out of place in that atmosphere, and we children couldn't wait to leave. But my father seemed very proud to show us all off, and this trip remains one of my strongest memories of him.

I have said that Die Urmama had two daughters and a son. That son, my great-uncle, Carl, changed his name from Cohn to Colbert. He was a Fabian Socialist and was probably the first to ignite the spark of social conscience in me. He founded a left-leaning daily paper, *Der Abend* (*The Evening*). Its maxim, printed on the masthead, was "Wo es Stärkere gibt, Immer auf der Seite der Schwächeren (In the presence of the powerful, always on the side of the weaker)."

His wife, my great-aunt Toni, had been a concert pianist. It was from her that I received the first initiation into my lifelong passion for music. I loved to go to their house on the outskirts of Vienna. In one of my earliest memories I am sitting under Aunt Toni's piano while she plays Beethoven. I had the feeling that the music was pouring over me. To this day I can conjure up the vivid memory of that waterfall of sound. I don't know how good a pianist Aunt Toni actually was, but my contact with her and with her music changed my life and enriched it enormously. Science became my profession, but music remains my religion.

My mother, who also loved music but played the piano only a little, saw to it that I got a good musical education. When I was eight years old, I started taking piano lessons and attended a music appreciation course in which a teacher told us moving stories about the famous composers. We learned about poor deaf Beethoven, and about Mozart, who died young and whose dog was the only mourner at his pauper's funeral. We heard about Schubert, who was often sick and who was only thirty-one when he died. (Mendelssohn seemed to have been the one composer who led a normal life. Indeed, it resembled our life, in a comfortable assimilated Jewish family.) After we had listened to these sad stories, we were asked to play something by the composer whose wretched biography we had just heard.

A few years later I attended another music course, which was designed to teach the rudiments of harmony. The teacher stood in front of a board and moved a large cardboard note around on a huge musical staff. Our task was to try to find the melody he was creating by singing the note as it moved up and down the lines. This was not an uplifting musical experience. For me the streetcar ride to these classes was the best part, because I was allowed to go by myself. I always stood near the driver and watched him while he drove, secretly waiting for my chance to be a hero. I had the unrealistic but exciting fantasy that one day the driver would faint and I would jump into his place behind the steering wheel, bring the car to a stop, and save the passengers from certain death. Needless to say, our trips were consistently uneventful.

For a time when I was a very young child, I became interested in technical things and told everyone I wanted to become a railroad engineer. Trains held a particular fascination for me. When I was about five years old, I was delighted one Christmas Eve to find a mechanical train running on tracks under the tree, a gift from my grandfather. My love for toy trains has persisted throughout my life. When our children were teenagers, I jokingly told them to marry early and to have children quickly so that I could buy them electric trains. On my fiftieth birthday they gave me a marvelous electric train set with two engines and many tracks and switches and cars of all sorts. The note pasted on the box said, "Why wait for the brats, enjoy it now."

In spite of our Jewish origins we celebrated Christmas, but not as a religious event. The tree with its lighted candles symbolized the return of the light in the midst of winter, and for us children, it marked an occasion for getting presents. Whereas we got toys and sweets at Christmas, the servants were given clothing such as sweaters, blouses, or socks—special gifts for them, but the sort of things we got as a matter of course whenever we needed them. Even as a small child I noticed this difference and wondered why it was so. My social conscience was already being formed.

At age six I began my formal education, but my first contact with school was a disaster. My mother thought I should go to the public school. On a practical level it was logical because the school was only a few minutes from our house. But more important, because she had progressive ideas, she wanted me to be with children of all sorts, not just upper-class ones.

On my first day the teacher made a little speech. "We are all equal in this school," he said. "Whether you are rich or poor, you are all the same. Even if your family owns a big corner house near the Ringstrasse, at this school you will be considered just another student." Unfortunately, that was an accurate description of the house my father owned and in which I lived. Although the teacher didn't look at me when he said this, and although his message was meant to make us all feel equal, his comment embarrassed me, and I felt that it set me apart from the others.

A few days later I experienced an even sharper sense of isolation. I was in the school bathroom, ready to go back to the classroom, when I realized that I didn't know how to put on my suspenders. At home there had always been someone to button them for me. I didn't know what to do, so I just sat there in the bathroom waiting in fear for help and for the punishment I knew I would soon receive. When the teacher arrived, after what seemed like hours, he was furious. I had to stand behind the blackboard at the front of the room where the whole class could witness my humiliation.

Of course, school also offered rewards. Children who did well got postcards featuring pictures of the emperor in various poses. But these small gifts were not what I recall from my early education. All the pleasant memories have been drowned out by the harsh, loud voice of my first teacher. I was not used to punishment, and I had certainly never been ridiculed at home. My parents believed that children should be loved and, whenever possible, praised. After a few weeks of school I was often sick in the morning or simply said that I was unable or unwilling to go to school. My parents soon recognized that the public school was not working out for me. They hired a woman who taught me at home for the rest of that year. To me, she seemed like the good fairy who rescued the children in my bedtime stories.

The following year I was enrolled in the Schwarzwald School, a progressive institution whose principal, Eugenia Schwarzwald, was an important figure in the community, like my great-grandmother, and a famous pioneer in progressive education. Most of the students came from well-to-do families like mine. We were under the care of three women, who used no threats, no punishments, and no pictures of the emperor—only a warm and playful atmosphere. From time to time Mrs. Schwarzwald came by like a demigoddess to see how things were going and to bestow a few kisses on some of her students.

As my school career began, the First World War broke out, and a wave of patriotism swept over Vienna. In school we sang patriotic songs directed against our enemies:

> Jeder Stoss, ein Franzos
> Jeder Tritt, ein Britt
> Und Die Flotte ist nicht Faul,
> Schlägt die Britten über's Maul.

> Every hit, a Frenchman
> Every kick, an Englishman
> And the fleet, no slouch,
> Hits the British in the mouth.

We crossed out the inscription *Made in England* on our toys and proudly carried the black and yellow flags of the monarchy. But the war was not real to our family until 1915, when my father, then a forty-two-year-old lieutenant in the reserves, was called into the army.

A few months later, as in normal times, my family moved to Altaussee for our summer holidays. One day when we were having tea in the garden and I was climbing on a wooden playhouse, I saw a telegram being delivered to my mother. She opened it and burst into tears. Then she read it aloud: "The Red Cross regrets to inform you that *Ober-Leutnant* Emil Weisskopf has been captured on the Russian Front and has been taken to a prisoner of war camp in Scobelev near Tashkent." My memory of the event is still vivid. Everyone in the garden began to cry—everyone except me. As a young child, I was very close to my father. I had often been told that I looked like him, and I know that I must have missed him. But I couldn't cry. Perhaps I was too young to grasp fully what had happened or what "prisoner of war" meant. In any case, I remember to this day the terrible feeling of guilt I had when my tears wouldn't come.

My mother was an extremely resourceful and strong woman. When she recovered from the initial shock of my father's predicament, she devised clever ways of getting news of the family and the war to him. She sent him telegrams written in a kind of code. Unfortunately, he was not always able to decipher her secret messages. Once after a great Austrian victory she sent him a telegram that read, "Greetings from <u>Gross</u>mutter <u>Sera</u>phine <u>Sieg</u>mund." The underlined syllables spelled out <u>Grosser Sieg</u> (great victory). My father couldn't figure it out. He sent back a message saying, "Who is Seraphine Siegmund?"

My father and his Austrian countrymen were lucky in their captivity. The camp was not far from Buchara. The emir of Buchara had been in Vienna as a student and spoke some German. Every month he invited the Austrian officers for dinner, for his entertainment and an opportunity for him to practice speaking German. After dinner, which the underfed prisoners looked forward to all month, there was a general discussion of Austrian music, literature, and culture, which made for a pleasant interlude for my father and his comrades.

During my father's absence, his law partner saw to it that we didn't suffer financially. Our life went on almost as usual. But my mother, now in sole charge of her three small children, must have felt overwhelmed. She hired a young man to act as a companion for her sons. His name was Hans Riehl, and he was an absolutely wonderful choice as far as I was concerned. Hans, who was in his early twenties, was interested in everything and knew a great deal about many subjects. He immediately sensed that my brother was drawn to poetry, history, and literature, so Hans helped him develop his interests in Western culture. I remember quite clearly that my brother once tried to build a Greek temple out of concrete in the yard of our summer house in Altaussee. Unfortunately, the columns always collapsed before the concrete was dry. I am convinced that, despite this failure, most of my brother's cultural interests were initiated by Hans Riehl.

Similarly, it did not take Hans long to see that I was a budding scientist. We spent our time together working on simple experiments and demonstrations. For example, he showed me the principle behind the electric bell. He built a model with an electric magnet and a lever that hit the bell. The current flowed from a contact through the lever. When the magnet attracted the lever, the contact was interrupted and the lever jumped back, hitting the bell. Then the current flowed again and the process was repeated. With the model and a drawing Hans Riehl described all of this to me clearly.

He also showed me how the lamps in the house were wired so that they always got the same voltage no matter how many were lit. He introduced me to the laws of planetary motion and explained that they are based on the same force that makes objects fall toward the earth. He gave me short popularized scientific books to read and encouraged me to ask questions. Like a gardener, he tended the first sprouts of my scientific interests.

And, to my joy, Hans was very musical and quite an accomplished pianist. He often took us to the opera. We saw Mozart's *Abduction from the Seraglio* and *The Magic Flute* and several other operas that contained the kind of spectacles or plots that children like. Before we went, he always told us the story of what we were going to see and played the famous arias on the piano.

12

As time went on we children were encouraged to attend concerts and became acquainted with all the important masterpieces, from Bach to Mahler. All this was taken very seriously: music was seen as a form of culture, not as entertainment. It was typical of the ambiance in our home that we never went to see an operetta, even though this was a Viennese specialty. It was considered below our family's dignity to indulge in such vulgar pleasures. To this day I am unfamiliar with the famous Viennese operettas, such as Lehar's *The Merry Widow* or *The Land of Smiles*. Only one operetta was tolerated: *Die Fledermaus,* by Johann Strauss, because it was the only operetta performed at the State Opera House. (We were not alone in such prejudice. The composer Alban Berg is said to have refused a blood transfusion in the Vienna hospital because he was afraid that it might contain the blood of a composer of operettas.) During the war our world continued to be a placid and stimulating place. We were happy children, sheltered and protected, living in our comfortable homes and enjoying the benefits of life in Vienna, even during wartime.

In 1917, the Russian Revolution broke out, and the military discipline in the prisoner-of-war (POW) camps collapsed. The Russians released all the POWs and left them to fend for themselves. My father and his friends were free to go home. But Russia was in turmoil, and these men faced difficulties in getting back to Vienna in the midst of a civil war. My father had learned some Russian in the camp, but not enough to get by on, so he and a friend decided to use their knowledge of Yiddish. They disguised themselves as Orthodox Jews and made their way from Turkestan to the West. It took them three difficult and dangerous months to reach the Austrian frontier.

There were no regular trains, so they jumped from one freight car to another. When they came to the Ukraine, they found that the region was in the hands of the rightist Petlyura group, famous for its violent persecution of the Jews. But after many close calls they managed to reach the German-Austrian armies and were finally brought to Cracow, which was at that time in Austrian hands. My mother, informed by the Red Cross, was there to bring my father home. He had grown a great black beard that scared us all a bit. He shaved it off immediately, though, and our life was soon back to normal, insofar as was possible during that period.

That first summer after the war we traveled by train to Altaussee as we had always done. The trip, a complicated affair with lots of luggage, resembled a migration. Our two maids always rode third class, whereas the family sat in the less crowded second-class cars, where the seats were soft.

Altaussee is a beautiful place, surrounded by mountains, green meadows, and forests. It borders a small, deep lake that reflects the mountains. This spectacle is especially beautiful at the Trisselwand, a straight, almost vertical expanse of white limestone rising twelve hundred meters above the water. Every mountain around the village has a distinct character—some are rocky, some wooded. Toward the south, between two of these mountains, you can see the Dachstein, a high range with glaciers and rocky peaks. For us children Altaussee epitomized natural beauty. The meadows with their ever-changing flowers; the forests with mossy rocks interspersed among huge pine trees, whose needles formed a soft brown floor where wild strawberries and blackberries grew; and the distant formations of the mountains remain for me the ideal of abundance and grandeur in nature. Even today, when I admire a mountain or a landscape, I realize that I am comparing it to Altaussee.

Altaussee was the summer home for many Viennese families, and a lively social life had been established, especially among the young people who came back year after year. A highlight of each summer was the weekly dances for the local population, to which we city kids were invited. We dressed up in the local costumes: *Lederhosen* (leather pants) for the boys and dirndl dresses for the girls. We learned to do the region's folk dances and songs in which only the boys took part. We sang folk melodies, as well as some slightly off-color variations in the local dialect. The dances took place at a rustic inn on the far end of the lake directly under the Trisselwand. From there we could see the white glaciers of the Dachstein reflected in the lake as the music reverberated from the nearby cliffs. Naturally, the summers also featured many love affairs, which added to the peculiar attraction of the place, especially when they were happy. When our hearts were broken, as they often were, we told ourselves we still had the lake, the mountains, and the beauty of nature for consolation. So the summers passed, and by the time I was a teenager, my life seemed set on a predictable course.

I don't recall suffering from many of the usual problems of puberty and adolescence. School was easy for me. Unlike many of my friends, I was good in mathematics. I didn't like history very much, but I was convinced that that was mainly my teacher's fault. My history teacher was virulently nationalistic, and I already considered myself a Socialist, more interested in the future of the world than in the exploits of the kings and emperors of the past. I must have been thirteen years old when the teacher reproached me in front of the whole class: "You don't even know a single date in history," he charged. I replied, perhaps because of my already keen interest in numbers, "That's not true. I know all the dates. I just don't know what happened on any of them." The teacher was incensed, although I had meant it only as a simple matter of fact. He muttered something that sounded like "Jewish impertinence" and summoned my mother to school for a conference. As usual, she managed to smooth it over, and since my marks were generally good, nothing came of this encounter.

Two or three times a month our family went to the theater. Like the opera, this was not seen as an amusement, although it gave us pleasure. It was considered part of our education. On the front of the Imperial Theatre in Copenhagen are engraved the words "Ei blot til Lyst (Not Only for Pleasure)," which represent the way theater was regarded throughout Europe.

I also went regularly to concerts. By the time I was twelve or thirteen, I was allowed to go by myself even in the evenings. My mother encouraged me in this. When I came home, she'd be waiting up for me to tell her about the concert while she made me scrambled eggs or a sandwich. I tried to describe every part of the concert, the music and the soloists. I gave my opinion of the performances, sometimes demonstrating the great climaxes on the piano. Our shared love of music established a great bond between my mother and me.

The performances of Bach's *St. Matthew Passion,* of Haydn's *Creation,* of Beethoven's *Missa Solemnis,* and of the requiems of Mozart and Brahms had another important effect on me. We received no religious education at home, and the classes in Jewish religious instruction that we had to attend at school were extremely uninspiring. I got acquainted with the meaning of religion and the concept of God through the liturgical masterpieces of the great composers (which I think is a much better way of providing religious inspiration than the classes offered by the schools).

Every day for eight years, on my way to school, I passed the Gothic-style Cathedral of St. Stephen in the center of the city. I did not always look at the wonderful medieval sculptures, but I must have been aware of them, and to this day I have a deep love for medieval art as one of the great expressions of European culture.

A more direct influence on me was the music lessons we all had to take. In families like ours every child learned to play some sort of musical instrument. My brother took up the violin, but he didn't play for long; and my sister started to study the piano, which she also soon gave up. But I loved the piano from the start. Part of the credit belongs to Aunt Toni's playing, of course. But there may have been another factor.

Carl and Toni were Socialists, and my relationship with them was, on some level, a rebellion against my father. He was politically conservative, and culturally he was a humanist whose interests lay in history, literature, and art. Neither music nor science interested him. By contrast, my love of music was only part of my relationship with my mother. I was secure in the knowledge that she would always encourage me in whatever I chose to do. It has been said that a boy who is his mother's favorite has an easier time in life. My very happy childhood and my adult life, which has been free of any serious psychological problems, may be due to the strength of my mother's affection for me.

We had a wonderful piano at home, a Boesendorfer grand, which my parents may have bought at the recommendation of Aunt Toni. But my playing got a serious boost only when Mr. Thornton became my teacher. I must have been thirteen or fourteen at the time. I suspect that Mr. Thornton was a failed concert pianist, but to me he was an impressive figure. He loved Beethoven and even imitated him by being often rather moody and stern. Our lessons were conducted in the shabby apartment he shared with his sister. His concert grand stood in the middle of a room whose only other furnishings were a piano bench and a chair or two for his pupils.

When I was assigned a new piece, Mr. Thornton would play it for me with great solemnity. Then he would say, "That's the way you will play this." Occasionally he would play something that was much too advanced for me only because I had asked to hear it. Once he played Beethoven's great *Pathétique* Sonata for me, and I was so moved that I asked him to teach me to play it. He replied that it was much too difficult for me. But the next time I came for a lesson I asked him again if I could try to learn the *Pathétique*. This time he said, "Well, if you really want to play it, really want to . . . well, yes, let's try it." I was too young to be able to play the *Pathétique* as Beethoven had intended it to be played, but Mr. Thornton did teach it to me, and I have been playing it ever since, improving, I trust, over the years.

When my grandson Colin, who takes piano lessons and loves to play, was fourteen, I played the *Pathétique* for him. He was so impressed by the great piece that he wanted to play it himself. Like Mr. Thornton sixty years earlier, I pointed out that it was surely too hard for Colin to play at this stage of his musical education. A few weeks later he came to my house and played the first movement with great verve and a few mistakes, probably just the way I had played it at his age. It gave me a sense of continuity, almost of immortality. I was tremendously moved, not only by his achievement but by a sense of the continuity of ideals and of the enduring value of great art. A few weeks later I bought him a Steinway grand to replace his mediocre upright.

Mr. Thornton also exhibited an interest in science, and since he knew that I wanted to be a scientist, he often asked me strange questions. Most of the time I couldn't answer them, they were so arcane. But our relationship, as pupil and teacher, deepened into friendship, and he not only taught me much of what I know about what music means but also contributed a great deal to one of the most important aspects of my life.

By now I was enrolled in the gymnasium, and I began several important friendships. Perhaps my closest friend was George Winter,

17

with whom I had mapped the progress of the shooting stars. But we shared more than an interest in science. He also loved music and was an ardent Socialist, and—a bit later—we both discovered girls.

There was a sharp division in the school between the intellectuals and the boys we called "the sport guys." (These days they are called jocks.) It goes without saying that George and I considered ourselves intellectuals. We looked down on people who were involved in such unimportant activities as sports, physical training, or calisthenics, and we reserved our most haughty disdain for spectator sports. We prided ourselves in being completely uninterested in the different soccer teams whose competitive fortunes were the topic of unending conversations among the sport guys. We did engage in two activities that could be called athletic: hiking and skiing. But we told ourselves that they were not really sports. They involved something much higher, the love of nature.

Our school was in a district where many middle- and upper-class Jewish families lived. The intellectuals were predominantly, though not exclusively, Jewish, and most of the sport guys were Gentiles. As intellectuals, George and I belonged to a discussion group that met for intense and serious conversations about literature and philosophy. We sat up late into the night exploring the fine points of books and arguing about ideas with the fervor only the young can muster.

Because I had skipped the fifth grade, George and I were in the same class, in spite of the fact that he was nearly two years my senior. At one point, in addition to all the other interests we had in common, we also loved the same girl, Annie Singer. She eventually married George, but for some time she and I were together a great deal. I was a late bloomer when it came to girls. My love for Annie was expressed in our common interests and in long conversations, not in making love. One Christmas I gave her something I wrote, a description in words of Beethoven's Ninth Symphony. Mercifully, this love offering has been lost.

As the youngest among my classmates, I spent some time listening to the older boys talk about their exploits with girls. I remember once we were all sitting in a circle playing a game in which you were asked whom you loved. You had to say my love begins with *A* or

*B* or whatever letter of the alphabet, and then the others would try to guess who it was. When it was my turn, I said, "My love begins with *C*," and no one could guess who it might be. I finally told them "centrifugal force" and everyone laughed. At that time—as most of my friends knew—I had had no real experience with girls at all. That came later.

All of this took place in the 1920s during a time of great sexual freedom in Europe. The mood during the years after the end of the war was very permissive. There were several similarities between those postwar years and the present. Drugs were widely used, although not nearly as widely as today, and there was also an epidemic of suicide among the young. However, I was completely untouched by all these troubles and led a very pleasant life. Since I didn't have love affairs at that time, I made them up for myself. Once when George and I spent a night in a cave after a hike in the country, I spun out a completely fabricated romantic life in order to impress him. I doubt that he was fooled.

My father had been much too busy to tell me anything about sex. My extremely proper mother was too embarrassed to talk about it, so I learned everything I knew from friends and from scientific books, which told me more about the reproductive habits of plants than about human beings. When I was sixteen I went to Paris for a visit. Before I left, my father took me aside and solemnly told me that Paris was a dangerous place for a young man and that I should not go with prostitutes. It had never crossed my mind to do such a thing. If I had gone with a prostitute, it would have been because he put the idea in my head. But the closest I came to love in the dangerous city of Paris was my infatuation with a statue by Maillol that I saw in the Louvre.

I had a great many other interests, including my still-developing concern about social justice. While I was in the gymnasium, one was expected to belong to a youth group of some sort. There were many choices: Boy Scouts, religious groups, Zionists, German nationalists, and Socialists. Communism didn't play a significant role in Austria, so there was no active Young Communist League. My brother joined Blau-Weiss (Blue-White), a group named for the Zionist colors. I was naturally drawn to the Socialists. I took part in the organization

19

of the Bund Sozialistischer Mittel Schüler (Federation of Socialist High School Students). There I met Bruno Kreisky, who became chancellor of Austria in 1970. We developed a warm friendship that lasted until his death in 1990, although for most of our adult lives we lived in different countries.

Basically, the activities of the youth groups were similar, whatever their ideology. We all met on the weekends and went on hikes, played handball (not my forte), and went skiing. Because the groups included both boys and girls, there was also some playing at love. But mainly we talked. In the evenings we'd stay up for hours in deep discussions about politics and philosophy. Sometimes we went to each other's houses or apartments, but we preferred to be outdoors sitting around a campfire. The various Socialist groups also met at larger gatherings at some historical or romantic spot like the ruin of a medieval castle. Sometimes we staged performances of historical events of "social significance," such as the peasant uprisings of the sixteenth century. Once in a while we had encounters with the rightist nationalist groups, and then there were real battles that resulted in bloody heads and injured limbs. But all of this only added to the romance of the thing.

Because we were young, we were also very critical of our elders, and one of our favorite activities was getting together to talk about the shortcomings of the Social Democratic party. The position of the party in Austria after the First World War was peculiar. Present-day Austria is the remnant of the great Austro-Hungarian Empire, a unique entity that embraced countries of very different nationalities, cultures, and degrees of development. When the monarchy fell apart after having been on the losing side of the First World War, most of these countries became independent entities—including Hungary, Czechoslovakia, Yugoslavia, and Poland—or were joined to existing neighboring countries such as Romania or the Soviet Union, often against the will of the population.

The German-speaking regions surrounding Vienna and west of the city became today's Austria and had a population of about 6.5 million, whereas the population under the Austro-Hungarian monarchy had been about ten times larger. This created a difficult situation. With almost two million inhabitants, Vienna, once the center of administration for the monarchy, became the capital of a small country. Most of the food for Vienna had come from the eastern provinces, which stopped supplying the city. The agriculture of the German areas was in shambles after the war. There was very little to eat in the city during the first few years after the war ended.

I remember those difficult years because our parents sent my brother, my sister, and me to live with a peasant family, whom they paid large sums of money. We didn't like being there, but at least we were reasonably well fed. During the summers of that period we could not go to Altaussee because of the difficulty of obtaining food. We rented rooms at a farm not far from Altaussee where the farmer was able to provide reasonable meals for us.

The Social Democratic party was led by many great personalities. Otto Bauer, Karl Seitz, and Karl Renner are three who come immediately to mind. All liberals opposed to bloody revolutions, they wanted to establish a state whose leaders were democratically elected, with a leftist majority that would organize the country and its economy according to Marxist principles. They favored protection of workers' rights, and they advocated state ownership of the factories. Although they wanted to provide a high degree of education for all citizens, they were particularly interested in giving these privileges to workers and peasants. These aims were subsumed under the term *Austro-Marxism*. The revolutionary character of this program had one good effect: it took the wind from the sails of the Communists, who never played more than a negligible role in Austria.

Although the Socialist leaders were very influential in the first provisional governments of the republic—Karl Renner was chancellor, and Michael Arthur Hainisch the first president—the Social Democratic party never succeeded in getting a majority in the elections, and the first Austrian governments were coalitions. Later on, the conservative party—the Christian Socialist party—formed the governments, mostly under the leadership of Jesuit priest Ignaz Seipel, a strong and cunning politician.

Austria was politically divided into the urban areas and the countryside. In particular, Vienna always had a strong Social Democratic majority. The Republic of Austria was then, as now, a federation of *Länder* with their own governments, roughly like the states in America. Vienna was one of these *Länder,* and the Social Democrats were the decisive party there. Indeed, their majority increased with each election. The party leaders decided to concentrate their reforms on Vienna, creating the so-called Red Vienna.

They succeeded in many respects, building a model city according to their ideas. Evidently, it was impossible to nationalize the factories, but they introduced a number of liberal, progressive innovations. They fought the critical housing shortage by constructing a large number of apartment houses in a modern style adjusted to the needs of workers and lower-middle-class families. These complexes had gardens, playgrounds, and community centers.

The Social Democrats also built new schools and introduced modern educational methods. Special emphasis was put on education for both the young and the old. The money for all this was acquired by extra taxes on property and on luxury items such as large apartments and private real estate. The housing developments and the adult education activities were the pride of Red Vienna, the product of the people and their optimistic attitudes.

Unfortunately, Red Vienna ended during the civil war in 1934, when the Social Democratic party was defeated and made illegal by the authoritarian rightist regime of Engelbert Dollfuss and Kurt von Schuschnigg. The dream was destroyed for good when Austria was taken over by the Nazis from 1938 until 1945. The party was forced underground, and many of its leaders were sent into concentration camps. Only after the Second World War, in the 1950s, did something like a new Red Vienna emerge. But it had none of the youthful vigor that had characterized it in the twenties.

Obviously, we young Socialists were enthusiastic about the development of Red Vienna. We criticized only the relatively slow pace of the developments, and we wanted more changes. We often expressed our dissatisfaction with the adults' handling of things. A number of us decided to do more than just talk. We composed a parody of the party leadership in a skit with words and music. Our leader was Victor Gruenbaum, who was famous for writing parodies. (He later changed his name to Gruen and became a successful and well-known architect in America, but in 1926 he wrote funny sketches and was sought after in many circles for this ability.) We met over a period of weeks, rewriting the lyrics to many of the popular Viennese songs. Fritz Jahoda was the leading musician, I was the second pianist, and the rest of our group were performers.

When we were ready, we had the temerity to invite the leadership of the Social Democratic party to a performance. Probably the only reason they came was that some of them had sons in our youth group. The politicians were flushed with success, having just gained the majority on the Vienna City Council. I suppose they expected some light ribbing but a general air of respect. What they got were songs like this one (which loses something in the translation):

> Zum Zeichen, dass wir die Mehrheit erreichten
> Lassen wir das Rathaus festlich beleuchten
> Was tun wir jetzt, was tun wir jetzt?
> Jetzt wird ein bisschen ausgesetzt
> Was tun wir dann, was tun wir dann?
> Dann fangen wir wieder von vorne an!

> To show that we are now in power
> We'll deck the city hall in lights
> And then what shall we do?
> Well, we need a little rest
> And after that, what comes next?
> Why, we'll start all over again.

At the time, our sketches seemed quite audacious and disrespectful. A scene remains vivid in my mind: There in the first row of the auditorium sat the great party leaders Otto Bauer, Julius Deutsch, Karl Renner, and the whole Social Democratic establishment. When we saw them, it occurred to us for the first time that they might find our skits insulting, and for a moment we were slightly worried. But to our amazement and joy the great men were obviously amused.

In fact, after our performance they asked us to write some more sketches in time for the national election. They wanted to use us in their forthcoming campaign. This time we were to have the Rightist party, then in power, as our "victims." Thus, our little amusement grew into a political cabaret. Gruenbaum wrote the lyrics to more

songs while the rest of us thought up jokes. We worked like a group
of TV comedy writers sitting together and spinning out ideas, and
we had a wonderful time. In the months before the election we took
our show to the provinces and to meetings in the city. We varied
the skits depending on the audience. In the city it was fast paced
with several jokes a minute for the sophisticated crowd. In the out-
lying regions we felt we needed more obvious jokes and more explanations.

Because the leader of the ruling party, Federal Chancellor Ignaz
Seipel, was a Jesuit, he was usually exempt from ridicule. One didn't
make fun of priests. But we couldn't resist. One member of our
group, Kurt Zinram, looked a little like Seipel, and when we dressed
him in a cassock with a clerical collar, he bore an uncanny resemblance to the chancellor. I played the *Aida* March, and Zinram/Seipel
strode onto the stage. The audience gasped and then broke out in
laughter and applause.

After one such performance the police wrote us a letter warning
us that we weren't allowed to make fun of the chancellor in this way.
"It's an offense against the state," they said. It was all right to have
fun, the letter went on, but we could not show Seipel in such a
ridiculous position. At first we were stymied. We tried to think of
what to do. Perhaps we should reply to the letter in the same offi-
cialese, which was in itself a kind of joke to us. But we decided that
the point would be lost on the police.

We came up with a response. At the next performance Gruenbaum
got up on the stage and slowly and somberly read the police letter
out loud. When he was through, I once again played the *Aida* March,
and Zinram, dressed in his cassock, came onto the stage. His head
was covered in a large box on which was written one word: *Censored.*
Even with the box on his head he looked like Seipel. It was an even
greater success than before.

Soon we were spending so much time at the cabaret that our
schoolwork suffered severely, but it had become more important to
us than anything we were doing in the classroom. Luckily, the director of our gymnasium was a Social Democrat, and he covered up
for us. While I was busy at the cabaret and enjoying it immensely
one aspect of my education continued undiminished. My interest in

physics and astronomy was growing. I was aware that I was living in one of the most exciting periods in the history of science. Quantum mechanics had been conceived and was developed from 1925 to 1927 by Niels Bohr, Werner Heisenberg, Erwin Schrödinger, Max Born, Wolfgang Pauli, P. A. M. Dirac, and others. They opened up the understanding of most atomic and molecular phenomena and laid the foundation of our knowledge about the properties of matter. (I could not have dreamt, while I was having fun at the political cabaret and beginning to study science, that in a few years I would be working with Born, Bohr, and Pauli.)

I realized that I did not yet have the training to have even the faintest idea of the meaning of the new concepts. In fact, very few people understood the new developments in physics when they were first presented. Nonetheless, I devoured all popular articles on the subject, both the good and the bad. Eventually several valuable books, particularly one by Hendrik Kramers and Hans Holst, helped me understand the basic ideas.

I was able to discuss some of these things with Franz Urbach, a friend who was a few years older than I and already had a postdoctoral position in physics at the university. He asked me to help him with his experiments, and I enthusiastically agreed to do so. This was one of the few periods in my life when I participated in experimental physics. We tried to do something that today is called nonlinear optics. We measured the optical properties of special crystals doped with metal atoms to try to find out whether the properties changed with the intensity of light. With the equipment available at that time we did not find any changes. Had we had the intensity of modern lasers at our disposal, we would have found interesting effects. Urbach was also just learning about quantum mechanics, but my contacts with him certainly deepened my understanding and interest in physics and in the philosophical implications of science.

In addition to his interest in science, Franz was a musician who led a choir specializing in old music. I joined the choir with as much enthusiasm as I had joined his experiments in the laboratory. When we got bored with the old composers, some of us formed a small choir of our own and sang the Brahms and Mozart requiems and other choral works of the nineteenth century. At this time I fell in love with Lisa Steiner, who was the girlfriend of Erwin Leuchter, the leader of our minichoir. She left Erwin for me, and we went out for two years until I left Vienna. We were drawn to each other by our common cultural and political interests.

I was eighteen and about to enter my first year at the university in Vienna when my father died suddenly. He was only fifty-three years old, but the hard years as a prisoner of war and his difficult escape from the camp had taken their toll. He was a classic example of a man obsessed by work. Every day by the time we got up, he would already have been working for several hours in his dressing gown. Then he went to the office and came home late. When I looked back on it, it seemed that he had spent relatively little time with our family.

My father's motto was simple: work, work, work. I have never been as driven as he was, but his manner of working has had a tremendous influence on me. He always said that work was the most important thing in his life. Moreover, he always maintained that as a Jew he had to work ten times as hard to reach the same position as a Gentile. The Austrian anti-Semitism of that time was effective but not deadly. Ostensibly, Jews enjoyed all civil rights; they just had a harder time succeeding. These circumstances seemed to have had a Darwinian effect on the Jews. Although there were many who could not succeed and who consequently led lives below the standards of Gentiles with the same abilities, those Jews who did achieve success were outstanding. During my father's lifetime a few of the many Austrian Jews who had attained prominence included Gustav Mahler, Sigmund Freud, Arthur Schnitzler, Max Reinhardt, Arnold Schoenberg, and many others. Some of Vienna's most gifted Jews enriched American culture after the advent of Hitler, and if my father was right, their achievements may have been partly due to the pressure of Austria's pre-Hitler anti-Semitism.

The structure and fate of the Austro-Hungarian monarchy had a lot to do with the large number of prominent Jews who flourished in post–First World War Austria. Many of the eastern provinces of the monarchy had large Jewish populations living under very difficult circumstances in ghettos or restricted areas. The more adventurous and intellectually active among them moved to the capital to escape the narrow, secluded life of the Jewish communities. When they arrived in Vienna, many of them quickly adjusted to the dominant German culture and neglected their Jewish religious traditions, perhaps because those traditions reminded them of the oppression they had escaped. Since they represented a group selected by a spirit of independence and adventure, many of them succeeded quickly in the new environment and were able to acquire influential positions. They were more deeply immersed in the German-Austrian culture than were many of the Viennese of longer residence. My family was an example of such assimilation, although my father observed more of the Jewish traditions than my mother, whose people had come from Hungary much earlier.

My friends and I definitely felt the dominance of the German cultural heritage when our parents and teachers encouraged us to admire Goethe and Schiller as well as Shakespeare, Balzac, and Dostoevsky, not to mention the works of the great German composers. *Bildung,* the extensive education based on Western culture with little emphasis on religion, played an important role in the lives of people like my family and their circle.

When I graduated from the gymnasium, I realized it was time to choose my life's work. Music had grown into a major preoccupation during the last two years of school when I had conducted the school orchestra in several memorable performances. We had been very successful in our program of waltzes for a Johann Strauss memorial. We had also performed Handel's *Concerto Grosso* and several other concerts that brought us acclaim and praise. I was tempted to leave science and go to the conservatory to train as an orchestra conductor. My mother had always wisely maintained that I would derive more pleasure from music if I remained an amateur, but for a while I seriously considered becoming a professional musician.

In the end, however, science won. When I had to choose a topic for my graduation paper—the so-called *Matura Arbeit*—which signaled the area of my future concentration, I chose "The Sun as a Source of Energy." My father had wanted me to be something practical like an engineer, but I knew that I wanted to work in physics. The University of Vienna, like other European universities, did not offer a general cultural or academic education. That had already been accomplished by the gymnasium. The university was where you received your training in a specialized field. It was similar to the American graduate school. It took three or four years and usually led to a doctorate in the field, but many students changed schools after their first two years. When I first attended the university, I lived at home so my life changed little. I continued my involvement with the Socialist youth group and our political cabaret.

However, I was also becoming very interested in the excellent two-year course in theoretical physics taught by Professor Hans Thirring, a rather well known physicist who was an outstanding teacher. The course concentrated on classical physics, the enormous edifice of knowledge that included an understanding of mechanics, electricity and magnetism, optics, and the theory of heat. Professor Thirring turned me in the direction of theoretical physics because I developed a deep curiosity about the intellectual underpinnings of physics, and I was given a glimpse into the unexpected, deeper relationships between natural phenomena.

It was in Professor Thirring's course that I first learned that magnetism, electricity, and optics are all manifestations of the same fundamental phenomenon, the electromagnetic field. And at the same time I began to understand that heat is a mechanical phenomenon, a measure of the random back-and-forth motion of the atoms or molecules in matter. Professor Thirring made me aware that many of the important insights into the physical behavior of matter, which were the main topics of his fascinating course, were developed in the nineteenth century. At the same time I became acquainted with other scientific discoveries of that period, such as the concept of atoms in chemistry and Darwin's theory of evolution. All this contributed to my admiration of the rich intellectual achievements of that fertile period.

During the nineteenth century, the industrial revolution changed the way people lived, worked, and traveled. Transportation by horse carriage was replaced by the railroad and later by the automobile and the airplane. The rights of workers, the eight-hour day, and medical assistance for the poor were recognized as necessary social innovations. The nineteenth century also brought about a revolution in the arts and literature. The art of the novel was developed by such great writers as Dickens, Dostoevsky, Tolstoy, Hardy, Balzac, Stendhal, and Manzoni, among many others, and the impressionist painters provided a new way of looking at nature. And, of special importance to me, some of the greatest music of all time was written during that period.

I have often thought that, given a choice, I would have wanted to live as a scientist in the nineteenth century. It was a time of opportunity and new beginnings in all the areas of culture. It was a time of great optimism for a better future. It was a time of innocence without the terrible knowledge we now have of all the negative consequences of much of what began at that time.

# TWO

⊠

# Becoming a Theoretical Physicist
# in Göttingen

WHEN MY FIRST TWO YEARS at Vienna University were nearly over, I asked Professor Thirring for guidance about how to continue my education. He advised me strongly to leave Vienna. "We can't teach you much more here," he said. "You should go to a better place, perhaps in Germany, where the new quantum mechanics is taught and pursued in laboratories." He suggested either Göttingen or Munich.

Quantum mechanics was and is one of the most revolutionary developments in the history of physics. Previous great developments had grown out of fundamental concepts not too different from those of our everyday experience such as particle, position, speed, mass, force, energy, etc. The world of atoms, however, could not be understood using those concepts alone. Surprising phenomena had been observed that demanded a different language for their description; new concepts were required to gain insight into the properties of atoms and their constituents. The French philosopher and mathematician Henri Poincaré said, "It is hardly necessary to point out how much quantum theory deviates from everything that one has imagined until now; it is without doubt the greatest and deepest revolution to which natural philosophy has been subjected since Newton."

I decided to continue my studies in Göttingen, but perhaps, in hindsight, it might have been better for me to go to Munich, where Professor Arnold Sommerfeld was the head of theoretical physics. He was addressed as "Herr Geheimrat," and he ran a very strict regime. He demanded that his students solve a large number of difficult problems in order to get a thorough grounding in theoretical physics. My knowledge and experience in advanced mathematics has always been weak, although I was good at math in high school, and I might have been able to achieve more had I gone to Munich and received the rigorous mathematical training Herr Geheimrat imposed on his students.

When I arrived in Göttingen in 1928, I found that while the regime was much less strict than in Munich, it had other advantages. The ranking professor of theoretical physics was Max Born, one of the leading contributors to the development of quantum mechanics. He kept somewhat aloof from his students and had a rather formal approach to physics. When lecturing, he tended to express everything in complex mathematical terms. While I preferred a less formalistic approach, I was nonetheless awed by his presence. I knew he was a great man, and despite all our differences, I wanted to emulate him. In fact, he was very friendly and encouraging to me. But when I came in contact with the new quantum mechanics, I felt that it was an esoteric theory far removed from ordinary human experience. I was still very young and idealistic, and my involvement with human affairs and social issues had become very important to me. Once again I thought I should change my choice of a career. I considered giving up abstract science in order to do something more directly important to human beings. I gave some thought to switching to medicine and becoming a physician.

After a year, I approached Max Born and expressed my fear that a life in science would isolate me from the concerns of humanity. He answered me with these prophetic words: "Stay in physics. You will see how deeply the new physics is involved with human affairs." He could not have known how right he was, in both a positive and a negative sense. Quantum mechanics revolutionized technology and led to modern electronics—transistors, semiconductors, and eventually computers. The new developments in nuclear physics would lead to the development of the atomic bomb. But neither of us could have known then that only twenty years later this would happen or that it would forever change our lives and the world.

In spite of several positive encounters with Born, my scientific collaboration with him was not as close as I had expected. He was, by nature, a man who kept to himself under ordinary circumstances. Shortly before my arrival in Göttingen, he had suffered a stroke and so was even less approachable. Still, we had some contact. One day Born called me to his office and told me he was preparing a textbook on optics. He knew that I had developed an interest in the behavior of light in crystals. He said he hadn't yet thought about the chapter on that subject and needed my help. I was to read the literature and to sketch out a synopsis of what I thought should be in the chapter. I wrote it up, but the result was so unlike Born's style that he couldn't include it in the book without making a great many changes. When the book was finally published, I was unable to recognize the chapter as something I had written.

The two senior professors in experimental physics were James Franck and Robert Pohl. I was especially attracted to Franck, who had won the Nobel Prize in 1925 for demonstrating that atoms in collision gain or lose energy in quantum steps. In contrast to the aloof Born, Franck was warm and very personable and loved interacting with his students, who were all impressed by him. He also had a group of lively young assistants who were very accessible to us and deeply devoted to their boss. It was a great pleasure to discuss physics with Franck. I was impressed by his intuitive understanding of science, his uncanny feeling for physics, and his marvelous way of working. He seemed able to predict with great accuracy the outcome of an experiment or the result of a theoretical calculation even when he was not well acquainted with the mathematical methods involved. Because of this we said he had a direct wire to God.

Göttingen's main professor of mathematics was Richard Courant, an extremely outgoing and gregarious person. He introduced the mathematical practicum to Göttingen. This unique concept involved study groups of two or three students who were assigned a mathematical problem to solve. It was a wonderful way to learn not only the solutions but the method of arriving at them. Although the teaching at Göttingen was not as strict and punctilious as it would have been under Herr Geheimrat in Munich, the atmosphere was challenging and the courses valuable. Courant also welcomed students to his home. What attracted me especially was that the house was almost always filled with music. Mrs. Courant was a wonderful pianist, and I spent hours listening to her play. Also, I was often invited to enjoy the chamber music that was the focus of many evenings at the Courant home. In the thirties the Courants moved to New York, where he founded the Courant Institute for Mathematics at New York University. (At this writing, Mrs. Courant, now in her nineties, lives in New York. I am told she still takes a cold shower every day and travels to Europe every year.)

The greatest mathematician teaching in Göttingen at that time was David Hilbert. When I got there, he was already in his late sixties and had long been a legend as one of the world's great mathematical minds. When he lectured, his assistant would write on the blackboard for him, since he could no longer manage to do that for himself. If he was not satisfied with the comments he heard in his seminars, he would say in his strong East German accent, "This is all chalk, chalk, chalk." Or he would mutter, "Crazy, crazy, crazy." Mostly, he was not satisfied with what he heard. He maintained that no lecture should last more than three quarters of an hour: a lecturer whose talk went on any longer was obviously using too many words. We all admired Hilbert and looked up to him, but we were more than a little in awe of him.

Another of our teachers, Edward Landau, was a well-known mathematician of the most abstract kind, so I had few academic dealings with him. But because he had two daughters, his house became a social gathering place, with frequent parties and dances. Since we were all young and had many interests in common, we had a wonderful time, and I remember vividly and with pleasure several parties where personal friendships developed informally.

I was attracted to Maria Göppert, called "Misi" by her friends. She was the daughter of a well-known professor of pediatrics, and her family belonged to the "good society" of Göttingen. She also studied theoretical physics, and her thesis was related to my own. For a time we were very fond of each other. Then a young American student came to the university. He impressed us by going to a car dealer, putting down a small pile of money, and driving out with a new car. In those days the idea of a student owning a car was preposterous. I don't believe it was the car alone, but pretty soon Misi was going out with the American, Joe Mayer, whom she later married. Misi was one of the few women who have received the Nobel Prize. She won it in 1963 for the work she did at the University of Chicago on nuclear shell structure, starting from a suggestion by Enrico Fermi. She shared half of the prize with the German physicist J. Hans D. Jensen, who had done the same work independently. The other half went to Eugene Wigner for his seminal work in quantum mechanics, which had little connection with Mayer's and Jensen's work.

In addition to the "great men" there were many younger teachers at Göttingen who were at the level of assistant professors in the United States. We learned a great deal from the junior faculty because they were much more accessible and we also saw them socially. I remember in particular Walter Heitler, Lothar Nordheim, and Gerhard Herzberg, who later settled in Canada and won the Nobel Prize for his work in molecular physics. Herzberg taught a course called "Introduction to Quantum Mechanics." This field was then so new that it was not yet part of most physics curricula, but Herzberg was young and eager to teach the latest developments. For me, Herzberg's course proved a great boon, because it helped me greatly to get acquainted with the new physics. The course was later published as a book that is still considered one of the best introductions to atomic physics.

The small group of students eager to understand quantum mechanics did not rely on Herzberg's course alone. We discussed it among ourselves, and we tried to read original papers. But we learned more from talking among ourselves and with these young professors than we did from reading the few published papers on the subject. Now I realize that the professors must also have benefited. Our conversations certainly clarified their thinking on this new subject of inquiry.

The new quantum mechanics was rather confusing to us. At the outset there were two seemingly disparate approaches. In July 1925, Werner Heisenberg published a seminal paper called "Quantum-Theoretical Reinterpretation of Kinematic and Mechanical Relations," in which he introduced concepts and mathematical methods rather foreign to the physical concepts of those days. A few months later Max Born and Pascual Jordan—and independently, P. A. M. Dirac—expanded Heisenberg's ideas and showed that a systematic theory of atomic behavior could be developed from them. These papers were very formalistic and are hard to understand even today.

The second approach was initiated by Erwin Schrödinger early in 1926 on a completely different basis. He made use of the ideas of Louis de Broglie, who had suggested in 1924 that electrons should exhibit wave properties. Schrödinger developed a wave mechanics for the electron that was much easier to understand and to work with than that of Heisenberg, Born, and Dirac. Later in 1926 Schrödinger and others showed that these two approaches were actually intimately connected. Both led to the same results, albeit by very different mathematical means.

As young scientists who had to learn all this, we were overwhelmed by the difficulty of both approaches, although Schrödinger's was easier to grasp by visualizing electrons in the atoms as wavelike vibrations. We were impressed by the fact that these vibrations correspond exactly to the known quantum states of the hydrogen atom. When I began to understand this, I experienced for one of the first times that joy of insight that was to play such a vital role in my life. But our older colleagues told us that to visualize these vibrations as waves could be misleading. The electron is neither a wave nor a particle. It exhibits wave properties under certain conditions—in particular, when it is within an atom—but it also exhibits particle properties, under different conditions. Quantum mechanics tells us when one or the other concept is applicable.* In many cases the wave picture is a great help for getting a better understanding of the quantum mechanical behavior of atoms, and we preferred the Schrödinger approach to that of Heisenberg. I am sure that the reader will understand the confusion in our heads when we began to study quantum mechanics in 1928. Some of the questions of interpretation that we faced then are still under discussion today.

When we were halfway toward an understanding of Schrödinger's waves, a new development confronted us. Dirac suggested in 1927

---

*I published a simple introductory essay on quantum mechanics in my book *The Privilege of Being a Physicist* (New York: W.H. Freeman and Co., 1989).

that one must introduce waves with four components in order to take care of the effects of relativity theory. This new "relativistic wave equation" created additional confusion not only in the heads of beginners like us but also in the minds of the more experienced people. Dirac's equation contained much more than a relativistic generalization of the Schrödinger equation. It indicated the existence of antimatter in a way that was not clear even to its author at first. Dirac said, in one of his talks, that his equation was much more intelligent than its inventor. It does happen from time to time that a scientist comes up with a new approach that leads much further than he or she had anticipated. Not until quantum field theory was developed in the thirties did we have a correct use for that equation. Much of my own future research was to deal with quantum field theory and Dirac's equation.

A very useful custom existed at Göttingen: Whenever a physicist from another place gave a talk at the colloquium, the graduate students had dinner with him afterward. This gave them a chance to get acquainted with the speaker and to ask him questions. In this connection I would like to report an event that may not really have taken place. It could be either one of those legends students concoct or an invention of my imagination, but it is worth relating because it could have happened.

A famous physicist, Abbé Lemaître, who also was a Catholic priest, gave a talk on the determination of the age of the earth by means of radioactive elements. Certain radioactive substances decay very slowly over a period of billions of years and gradually transform into other substances. One can calculate the age of the earth from the amount of these so-called daughter substances found near the decaying materials. Abbé Lemaître told us that such investigations had revealed that the earth was about 4.5 billion years old.

When we were sitting with Lemaître after his talk, someone asked him whether he believed in the Bible. He said, "Yes, every word is true." But, we continued, how could he tell us the earth is 4.5 billion years old, if the Bible says that it is about 5,800 years old? He said, I suppose with tongue in cheek, "That is no contradiction." "How come?" we nearly shouted. He explained that God made the earth 5,800 years ago with all the radioactive substances, the fossils, and other indications of an older age. He did this to tempt humankind and to test its belief in the Bible. Then we asked, "Why are you so interested in finding out the age of the earth if it is not the actual age?" And he answered, "Just to convince myself that God did not make a single mistake."

The distinction between theoretical physicists like Born and experimental physicists like Franck became more pronounced in the twentieth century than it had been before. In the past physicists performed experiments to discover new phenomena and to learn about the quantitative behavior of nature. At the same time they tried to understand what fundamental laws of nature could be deduced from the results of their experiments. Over time, the techniques of experimentation, as well as the mathematical tools for the analysis of results, became so sophisticated that a kind of division of labor between theoretical and experimental physics became necessary. Very few modern physicists could excel at both tasks. Two who could were Enrico Fermi and Felix Bloch.

For me it was exciting to penetrate deeper into quantum mechanics and to begin to understand why atoms and molecules behave as they do and how they assemble to form the objects that surround us in nature. After a long period of searching and misunderstanding, what I was looking for suddenly became clear and transparent, and I experienced once again the joy of insight. Now it was intensified as I penetrated the new theories that had been formulated only a few years earlier. They were so completely different from the approaches of the past and shed so much new light on the old concepts of observation, causality, and relativity that it was like the opening of a new world. I can best describe the joy of insight as a feeling of aesthetic pleasure. It kept alive my belief in humankind at a time when the world was headed for catastrophe. The great creations of the human mind in both art and science helped soften the despair I was beginning to feel when I experienced the political changes that were taking place in Europe and recognized the growing threat of war.

During the 1960s I tried to recall my emotions of those days for the students who came to me during the protests against the Vietnam

War. This, and other political issues, preoccupied them, and they told me that they found it impossible to concentrate on problems of theoretical physics when so much was at stake for the country and for humanity. I tried to convince them—not too successfully—that especially in difficult times it was important to remain aware of the great enduring achievements in science and in other fields in order to remain sane and preserve a belief in the future. Apart from these great contributions to civilization, humankind offers rather little to support that faith.

When I entered the university, I did not know what branch of physics I would concentrate on. My work with Franz Urbach in Vienna made me lean toward experiments, and my admiration for James Franck had strengthened that idea. On the other hand, the inspiration I had received from Professor Thirring's lectures and my burning interest in the basic questions raised by the new quantum mechanics drove me to theory despite my lack of interest in complicated mathematics.

A small incident may have been decisive. I took a laboratory course in which students were supposed to perform experiments. My immediate supervisor was Hertha Sponer, a rather abrasive woman. One day I had the misfortune of breaking an important part of an instrument, and Miss Sponer said, "You will never become an experimenter." Although her judgment seemed hasty, I took her word for it at the time, and her prediction proved true. Because of this incident I would have had difficulty in getting accepted to do a doctor's thesis in experimental physics. I do not regret the choice I made. It left me more time to think about the fundamentals and about the philosophy of our science. My aversion to complex mathematics may have been an advantage in the long run. It forced me to think harder about the essential ideas and connections that are sometimes hidden in mathematical formalism.

One of the most decisive influences on my scientific outlook came from Professor Paul Ehrenfest. He had been born and brought up in Vienna and taught at the University of Leiden in Holland. In 1929, when Born suffered his stroke, Ehrenfest came to Göttingen to teach for a year, and I was lucky enough to meet him. Ehrenfest took a special interest in me because I came from Vienna, and we spent a lot of time together. As a theoretical physicist, he also shunned long mathematical derivations, a trait that aroused my enthusiastic approval. He felt that mathematics often covered up essential points. At one of our first meetings he said to me, "Don't let yourself be impressed with all the formalism you will see here in Göttingen. Don't be awed by all these great scientists. Ask questions. Don't be afraid to appear stupid. The stupid questions are usually the best and the hardest to answer. They force the speaker to think about the basic problem. Try to get to the fundamental ideas, freed from the mathematical thicket. Physics is simple but subtle." This last remark made the greatest impression on me and it has been one of the guidelines of my thinking.

Ehrenfest was always in the midst of discussions about the latest developments in quantum mechanics. Everyone admired his determination to get to the essential point, and we liked him for asking the pertinent, contentious questions that other people had been avoiding. He loved young people, and teaching was obviously his greatest pleasure. But he was also a deeply troubled man who had periods of severe depression. He was often moody and sometimes impatient. His life ended tragically in suicide. When he was depressed, he used to say that physics was too difficult for him, that he couldn't follow it. But the cause of his suicide must surely have been more complicated than his frustration with science. The fate of his two sons might have been a contributing factor. One died in a mountain-climbing accident. The other was mentally deficient, and his father killed him before he shot himself.

Ehrenfest's daughter, Tanya, was a completely original person. Like one of her brothers, she had been educated at home by her parents, who opposed the way schools were run. They wanted to encourage their children to express their individuality, and they certainly succeeded with Tanya. She dressed in a sort of sack, never wore makeup, and was completely different from Misi, who was a conventionally dressed, good-looking girl. For Tanya there was no such thing as a "nice" dress. I was fascinated by her.

Göttingen was important not only to my development and training as a scientist but also to my maturing as a man. This was the first time I had been away from home. The university community became my world, and I felt the influence of both its bad and its good aspects. A negative factor was the dominance of the *Studenten Verbindungen* (student organizations or fraternities), which were nationalistic— many of them anti-Semitic—and were typical examples of the growing Nazi movement that came to power a few years later. The members of these student groups spent their free time drinking beer and roaming the streets, where they could be heard howling late at night.

On the other hand, there was also a large group of students and young professors who were intellectuals. The two groups formed a dichotomy similar to the sport guys–versus–intellectuals schism at the gymnasium in Vienna. My circle of friends was made up almost entirely of scientists, but we were involved in the whole range of interests the world had to offer. We cared about music, art, and theater. But politics was the most compelling subject in the late 1920s; you could hardly escape it. The economic crisis was just beginning in Germany, and the fraternities were not the only place where we sensed the beginnings of the Nazi movement.

In Göttingen my political interest took a less active form than it had with my left-leaning cabaret friends in Vienna. Now my friends and I were observers rather than participants in political life. The German political situation was more serious and had greater consequences, weightier implications, and more potentially dangerous conflicts than the relatively parochial world of Viennese politics. On one side were the increasingly powerful Nazis, and on the other, the influence of the strong Communist party. Everything in Germany was more intense: Political events were exciting and threatening, and life was richer than it had been in Vienna. The lively spirit in liberal circles led to all sorts of political and cultural events every day.

European university life at that time bore little resemblance to its American counterpart. We weren't required to go to classes, and in most fields, except medicine and law, there were practically no exams and no student advisors. Ninety percent of the students were just average kids who did as little academic work as they could and spent most of their time drinking or carousing with the fraternities. Our circle felt superior to these students, although we were certainly not always serious. When not studying, we had lots of fun.

Music remained a big part of my life. I played the piano with several chamber music groups, and I accompanied a fellow who lived in my boarding house and was the tenor at the local opera company. The opera house where he sang was typical of the sort of places that exist, along with theaters, in even the tiniest German town. These provincial playhouses weren't as good as the ones in the big cities, but they provided cultural outlets and fostered an interest in music and theater all over the country.

My friendship with the music critic on the local paper got me work as a reporter and I was sent to a week-long Brahms festival being held in Jena under the baton of Wilhelm Furtwängler. We all admired him extravagantly; I had gone to many of his concerts in Vienna, where he often conducted the Vienna Philharmonic. His Beethoven and Brahms interpretations were famous for their dramatic climaxes. As a reporter, I went to all the concerts and rehearsals. I still remember with pleasure the week I spent listening to Brahms chamber music, the symphonies, and his great *Requiem*. The stories I wrote for the paper seemed a small price to pay for the pleasure the music had given me.

These pleasant distractions did not, however, obscure my main reason for being in Göttingen: to prepare myself for a life in science. In 1929 the time had come to choose the subject for my doctoral thesis. In American and European universities in those days, either a professor designated a student's Ph.D. thesis topic or the professor and student chose one together after lengthy discussion. (A story about Max Planck was widely told in those days when the subject of a thesis came up. A student came to him and asked for a problem to solve for his Ph.D. thesis. Planck is said to have answered, "If I knew of a solvable problem, I would solve it myself.") I should have been talking to Born about my thesis topic, but since he was still incapacitated by his stroke, I decided on my own to investigate the way atoms emit light. This subject interested me greatly, and it was something I had been involved with for some time.

In the course of my investigations I was caught by the following problem: One of the main discoveries stemming from quantum theory was the existence of specific quantum states in the atom. An atom can assume a series of discrete states with distinct energies. The ensemble of these states is called the spectrum of the atom, and the state with the lowest energy is called that atom's ground state. When the atom is brought into a state of higher energy—an excited state— it subsequently performs a transition to a lower state, and the energy difference is generally emitted in the form of a light quantum. The radiation of light by an atom always results from such a transition, from a quantum state of higher energy to one of lower energy. To give an example in practical terms: the current in an electric light bulb excites the atoms into higher states, and their return to ground state produces the light.

There is a fundamental relation between the energy of a light quantum and the frequency of the corresponding light. Thus, the frequencies of light emitted by an atom would seem to be well determined, or "infinitely sharp," because they must correspond to the energy difference of the two states between which the transition occurred. In fact, however, the frequencies of light emitted by an atom are not infinitely sharp, for a number of reasons including the motion of the atom and collisions with other atoms. But we were interested in the lack of sharpness for isolated atoms at rest.

In 1927, P. A. M. Dirac had published a seminal paper called "The Quantum Theory of the Emission and Absorption of Radiation." This paper opened up a new field called quantum field theory, which allowed one to calculate systematically the rate at which atoms emit or absorb light. But the new methods did not permit the determination of the frequency distribution—that is, the deviation from ideal sharpness. I found a way to get at the distribution for the simplest case: the emission of light in a transition from the first excited state to the ground state of the atom. Thus, I chose this problem as the topic of my thesis.

The answer I found was what everybody expected: it was the frequency distribution of the radiation from an ordinary radio antenna. When I wanted to apply my method to light emitted by a transition between any two excited atomic levels, however, I ran into mathematical difficulties. Because Born was unavailable, I discussed my problem with Eugene Wigner, a Hungarian physicist who worked in Berlin but made frequent trips to Göttingen. Although only a few years older than I, Wigner was far ahead of me in knowledge and experience.

I had worked out the methodology, but I wasn't able to carry out the application of my idea to the actual calculation. Wigner found my idea interesting and helped me enormously as we worked out the solution. He and I together published the results: Radiation emitted in a transition between two excited atomic levels was different from the frequency distribution of a radio emitter of the same frequency. (The only person not surprised by this was Hans Franck, who as usual predicted the result, no doubt due to his famous direct line to God.) Because it was a collaborative effort, I couldn't use the work as my Ph.D. thesis. Instead, my thesis applied the same methods to the problem of the reemission of light absorbed by atoms.

The method Wigner and I devised to solve our problem had special significance for the development of field theory. It was the first time that solutions were found without the use of "perturbation theory," that is, without assuming that the interaction between the atom and the light is very small. Instead, we assumed that the influence of all those atomic states not directly involved in the emission of radiation could be neglected. The so-called Weisskopf-Wigner method of solving problems of quantum field theory was later applied to several other problems.

Wigner occasionally had an unusual approach to aspects of his work. Once he showed me a paper he had written in which he said that a certain formula had been derived by Gregory Breit. He gave the citation, with the name of the paper, the page, and so on. When I looked up Breit's paper, it turned out Breit had not derived the formula. I mentioned this to Wigner, who said, "Oh, yes. But Breit could have done it in that paper if he had thought of it." Another time I showed him a paper that, from its title seemed to have some relevance to our work. Wigner looked at the title, thought about what it could mean, and then worked out the solution to the problem. I said that I would like to read the paper. Wigner said, "Don't bother. Either it's what I just wrote down and you don't have to read it because I've showed you the solution, or it's not, so the man who wrote it is wrong and it's not worth reading."

I was unusually lucky in my joint publication with Wigner, since his name is one of the few that follows mine alphabetically and I thus appeared to be the prime author. I had made a vow that when I collaborated on a paper I would always put the names of the authors in alphabetical order in order to avoid discussions about this subject. As an aside, many years later when I wrote a textbook on nuclear physics with a young collaborator, John Blatt, I insisted that my name be second. The publisher opposed this, since my name was much better known. He said selling the book was more important than a vow. I tried to convince him by this argument: When you say "Weisskopf and Blatt," the emphasis is on "Blatt," whereas in "Blatt and Weisskopf" the emphasis is on "Weisskopf." To my amazement he finally agreed.

Eventually the time for my oral Ph.D. exam arrived. Sixty years have passed since I walked into that room in Göttingen where three professors waited to ask me the questions that would test my knowledge in my field and give me the seal of my profession, but I remember it as if it had happened a month ago. There sat Born, Franck, and a distinguished physical chemist named Arnold Eucken. I was able to answer Born's questions because I had guessed correctly what he might ask. Franck caught me by surprise when he said, "I understand you were at a meeting of the German Physical Society last week. Can you give us a report about it?" I told him what I remembered about the meeting, and he had no further questions.

With Eucken I expected to have some trouble because I hadn't done much work in chemistry, a field in which I was only marginally interested. But I was familiar with Eucken's textbook on the subject, so I had studied it fairly thoroughly. Unfortunately, I had skipped the paragraphs in small print. Of course, he asked me a question from one of those paragraphs, and I didn't know the answer. Fortunately, he also raised questions from other sections of the book, so I squeezed through. I had my degree.

Because Göttingen was not far from Berlin, I often went there, especially when I needed to see Wigner about our common work. I usually visited the apartment of Eva Striker in the center of the city. Later, as Eva Zeisel, she became a well-known potter, but even then she was on her way to fame. She maintained the kind of perpetual open house that was common in those days—you could simply come and go as you pleased. Often I stayed there overnight. I met many interesting people at Eva's, including the physicist Leo Szilard, the writer Manes Sperber, and other people connected with the leftist anti-Nazi movement.

One evening Eva took me aside after we'd been out for a meal with a large group. "Viki, you are really terrible," she said. "You never pick up the check, and I know that you have plenty of money from your family. It just doesn't go with your character for you to be so stingy." I had never given this subject much thought. My mother was a rather economical woman, and I must have inherited that tendency from her. I was a bit taken aback by Eva's remark, but I was really grateful to her. It caused a fundamental change in my attitude toward money. Recently I visited Eva and her husband, Hans Zeisel, in Chicago. When I told him that story, he said, "Oh, I wish you had influenced her instead of her influencing you."

Of course, Berlin was culturally richer and much more exciting than Göttingen. We went to Max Reinhart's theater, to the extraordinary Berlin Opera, and to the political cabarets that were the rage in those days. The cultural climate was still under the influence of *The Threepenny Opera* by Bertolt Brecht and Kurt Weill. With its glorification of the seamier side of society, *Threepenny* created a style that set the tone and gave color and character to the cultural ambience in Germany during the years from 1924 to 1932. In 1933 the Nazis got rid of all that.

The *Threepenny* spirit was a warm embrace of the underworld, where people were not only poor and involved in the shady side of life but also resourceful and exuberant. Mackie Messer, the *Threepenny* hero; Polly, his girlfriend; and Jenny, the pirate girl, were exciting, lively people with attractive, sensuous natures in spite of their irregular way of life. There was a good deal of sentimental poetry mixed in. The musicals and plays of Brecht and Weill and their followers, such as Hans Eissler, Paul Dessau, and others, were very popular with the large crowd of young, slightly leftist people who longed to be free of bourgeois prejudices about society, sex, and the way they conducted their lives.

On a higher level the old prejudices were attacked by poets such as Georg Trakl and by political writers such as Vienna's Karl Kraus and Berlin's Kurt Tucholsky, who was later murdered by the Nazis. Kraus was our hero because he had been fighting injustice and dishonesty by the military, the civil authorities, and the press since the old days of Kaiser Franz-Joseph. Kraus died shortly before the Nazis came to Vienna, but he lived long enough to see Hitler come to power in Germany. He asked his friends, "Tell me, why did I actually object to the kaiser?"

I must admit that when I got to Berlin in those days of change and emerging liberation from the old mores, I was just a little shocked. The sexual revolution in Berlin in the twenties was particularly striking for me. I was young, somewhat prudish, and certainly a bit provincial. I wasn't interested in participating in orgies. My love affairs—by then there had been several—were serious, important, and intense. Some of my friends boasted of having affairs with two or three girls at the same time, but I remained steadfastly romantic.

My friends and I recognized that the lively parties and late nights were a backdrop to a much more important drama: the political developments in Berlin of the twenties and thirties. They soon touched our lives. Manes Sperber and I had become friends. (Later, when he was a well-known writer in America, we saw each other again.) One day, in Berlin, he told me he needed a passport to get someone out of the country. Manes was a Communist, and it was getting increasingly dangerous to be a Communist in Germany. He wanted to give my passport to his friend. I did not like the idea.

He persisted, claiming it would be easy: All I had to do was report to the Austrian embassy that I had lost my passport. They would immediately issue me a new one. It was a very difficult decision for me, but in the end I couldn't bring myself to do it. Although I felt bad about it, I refused. Manes, who was young and passionate, bawled me out. He said that I was just "one of the bourgeoisie." Perhaps he was right. In Austria my political involvement had been mainly in our little cabaret group, performing our harmless satires about political figures. But this was real life; in fact, it was life and death. I asked myself why I was behaving this way when morally and politically I sided with Manes. He felt he had to fight and act on his beliefs; at that point in my life, I didn't. Science and music were still more important to me than political ideals. There would come a time when the issues of science and social conscience would come together for me, but I was not yet ready to risk everything for a political cause. I was still more involved with my future and my education.

# THREE

⊠

# Near the Founders of Quantum Mechanics

IN APRIL 1931, after I received my Ph.D., I went home for the summer. I had to decide what to do next. It was the middle of the Great Depression, and since there was no chance that I would get a job anywhere, I decided to do some sort of postdoctoral work. Today such jobs include a salary, but I had to be supported by my family. In order to learn as much as possible, I wanted to get a position with a well-known physicist. I went to Leipzig to work with Werner Heisenberg.

Heisenberg, one of the founders of quantum mechanics, had a special intuitive way of getting to essential points. This, together with his incredible persistence and determination, made him the most prolific and successful physicist of his generation. Whenever important problems turned up during the development of quantum mechanics, it was usually Heisenberg who provided the solutions. By conceiving new ways of looking at situations, he charted the course of further research.

His most widely known contribution is the Heisenberg uncertainty principle, which formed the basis of a new understanding of atomic reality. It defines the limit of the scientific applicability of ordinary concepts taken from everyday experience, such as position and velocity of a particle. In many instances, Heisenberg showed, these terms could not be applied in the same unambiguous way as was thought before the advent of quantum mechanics. While classical concepts are certainly good enough for dealing with macroscopic objects, they are only of limited use for atomic or molecular dimensions. A new language had to be found in order to deal with atomic reality, in which the old concepts play a different role and have only limited validity. If Heisenberg's principles had been called "limiting relations," much philosophical discussion about some intrinsic uncertainty in nature could have been avoided.

Heisenberg, thirty-one that year, had already won the Nobel Prize. It had been six years since he had laid the groundwork for quantum mechanics and formulated the uncertainty principle. He had long been a great scientific hero of mine, so I was delighted at the prospect of working with him.

Because Heisenberg was not yet married, he spent a lot of time with his students. Among other things, he was a great Ping-Pong player. He had one Japanese co-worker, Yoshio Nishina, who played better than he and who beat him regularly. I recall one occasion when Heisenberg, who was not a good loser, disappeared for three days after a defeat by Nishina. Heisenberg was also an amateur pianist. I still recall how he played Beethoven's extremely difficult Hammerklavier Sonata for us. His performance was technically perfect but almost completely devoid of passion.

Once again I found a wonderful group of young people gathered around a great teacher. Among these were Felix Bloch and George Placzek, who became my close friends. Bloch was squarely built and tall, and I always thought that his character was also squarely, heavily built and large. An incident that took place some years later when we were both in Copenhagen is typical of the way Bloch functioned. A group of us planned a somewhat harebrained practical joke: we would publish our best papers under a single invented name. There was a good chance, we thought, that this fictitious character would get the Nobel Prize, and we would then share the money. (There was actually a precedent for this crazy idea. A group of mathematicians had invented a certain "Burbaki" and pulled a similar prank. They published some of their best works in a book that was supposedly written by him. Burbaki became a "famous mathematician," but there were no Nobel prizes for mathematics.) Everyone in our group was ready to sign an agreement to proceed on this foolish venture. Only Bloch refused, and he was the only one of us who eventually did get the Nobel Prize. After a while we dropped the idea.

From the point of view of physics, 1932 was a very good year. Many new things were being discovered about radioactivity and nuclear structure, although scientific investigation in these areas was still in its infancy. The most exciting event of the period was the theoretical prediction and subsequent discovery of antimatter. In 1931, Dirac had predicted the existence of a particle identical to the electron but with an opposite charge. He called it the "anti-electron." Dirac developed the idea from the principles of quantum mechanics and the theory of relativity—not without some influence from J. Robert Oppenheimer and Igor Tamm. A year later in the United States, Carl Anderson independently discovered the predicted particle in cosmic rays. He called it a "positron." His discovery was shortly thereafter confirmed by Patrick Blackett and Giuseppe Occhialini in England.

One problem we discussed in Leipzig was the question of why an electron could come out of a nucleus, as observed in radioactivity, when it was impossible for an electron of the observed energy to exist within a volume as small as that of a nucleus. One day we were sitting in a coffeehouse outside a sports building with a swimming pool, discussing this problem, when Heisenberg said, "Look at how people go in and come out of that building all dressed. Do you conclude that they were dressed in the pool?" As it turned out, the electron does not come from inside the nucleus but is created near the nucleus.

I had planned to stay in Leipzig for a year, but around Christmas I got a letter from Erwin Schrödinger, who was then a professor at the University of Berlin, asking me to come and be his assistant. This was a paid job, something like the position of research assistant in the United States. I was young and inexperienced, and I knew the job would entail simply doing what Schrödinger told me to do. The job was offered to me for only one semester, after which his regular assistant, Fritz London, was expected back from the United States. But I quickly accepted. One didn't turn down a chance to work with Schrödinger, no matter how lowly the position.

Besides, the idea of being in Berlin appealed to me greatly because of the many famous physicists there, such as Albert Einstein, Max

Planck, Walther Nernst, and Leo Szilard. I remember vividly the weekly colloquia with the great men all sitting in the first row. Once Hans Kopfermann, who became a close friend later on, reported on different experiments to determine the value of Planck's constant, referred to as the quantum of action. All these experiments had resulted in the same value of the constant. In the discussion, Nernst asked Kopfermann a few questions. He was known as a skeptic about quantum theory even then, in 1932, when it was generally accepted. After Kopfermann answered his questions satisfactorily, Nernst loudly said to Max Planck, sitting next to him, "Well, my colleague, then you still have a few chances to be right."

Schrödinger, unlike Heisenberg, worked alone. He wasn't the type to go out with a group of students and encourage us to talk about the things we were working on. I didn't see much of him outside of the university. My duties were extremely light: I corrected the homework he assigned the students. The rest of my time was my own, and I used it to pursue my own line of research.

I was interested in the general question of the frequency of atomic radiation, but not only that for isolated atoms at rest, which Wigner and I had calculated—the so-called natural line width. I wanted to know what other effects broadened spectral lines. There is a motion of the atoms that causes what is called a Doppler shift: this is similar to the change in pitch of a siren on a moving fire engine. The interaction between neighboring atoms also changes the energy levels of the atoms and, consequently, the frequency of the radiation. Collisions between atoms interrupt the radiation and cause a broadening of the frequency of that radiation. I studied these effects and published a review article on the subject at the end of 1932.

By 1932 bands of young Nazis freely roamed the streets of Berlin. My office, facing the courtyard of the university, provided a front-row seat to what was happening. Often I saw the Nazi gangs who regularly searched out and beat up Jewish students or those who looked Jewish. The police didn't interfere, and more than once I had to pull one of the boys under attack into my office so that he could escape through the back door. As an employee of the university, I had a pass that let me through the police lines, which were by then set up around the university, causing the place to look like a fortress under siege. However, the police were completely ineffective in stopping the turmoil—or perhaps were unwilling to try.

I rented a room from a man named Teschemacher who had three daughters. They all studied at the nearby Academy of Arts and attracted a group of interesting people—artists and musicians. Through them, I was in touch with the lively, exciting, and somewhat hysterical world of the avant-garde, a group of people strongly opposed to the Nazis, who had branded these artists as purveyors of decadent culture.

At the Institute of Technology, I met Fritz Houtermans, a young experimental physicist educated in Vienna and Göttingen who was, like me, an assistant to a senior scientist. Houtermans was an amusing man, full of jokes and fun, and he had many interesting ideas for experiments. He and his wife often invited colleagues over for informal evening gatherings, which they called *Eine Kleine Nacht Physik*. On these "little physics nights," we would drink heavily and engage in lively discussions, often until the early morning. The topics ranged from music to science to politics, but our main focus was always physics.

In the early spring it was carnival season, and we all went to the gala ball at the Academy of Arts organized by the avant-garde artists. Invariably, these evenings degenerated into orgies, which I found distasteful. Not unexpectedly, the Nazis used such occasions for random violence. A group of them crashed one party I attended and began beating people up. The party ended in horrible confusion.

I also saw a lot of Eva Striker during that spring, and I had many occasions for meeting people outside of the scientific community. Among my new friends was a beautiful, tall, black-haired woman named Ruth Benario who was a professional photographer. Soon Ruth and I were going everywhere together.

When it was time for his assistant to come back from America, Schrödinger asked me what I was going to do next. I hadn't any idea, but Schrödinger did: he would arrange a Rockefeller fellowship for me. The European Rockefellers were a great distinction—everyone wanted one. You were given what seemed like an impossibly generous sum of money—$150 a month—and you could pick where you wanted to go to work or study. There was no special reason for Schrödinger to do this for me, and I was very grateful to him. In a short time I received a letter from Warren Weaver, director of the Rockefeller Foundation, informing me that a grant had been set aside for me. However, it wouldn't start until the fall, which was still more than half a year away.

Since I had no other plans and I had always been deeply interested in what was going on in the Soviet Union, I thought that this would be the perfect time to go there and have a look. I had talked to people who had been there. Some had raved about how wonderful it was; others had told terrible stories. I wanted to see for myself. This was my chance to see communism in action. Several of my friends had gone to the Soviet Union to live. One of them, Alex Weissberg, was the manager of the Physico-Technical Institute in Kharkov, where he was very successful and seemed to be very happy. We had known each other in Vienna when Alex was the only Communist in our Social Democrat group. When I wrote him to say that I was thinking of visiting Russia, he offered me free room and board and encouraged me to come.

I had asked Ruth Benario to come with me. In May we traveled by train to Warsaw, where we wandered like tourists around the city for a day. Our next stop was Leningrad, where we saw the great impressionist collection at the Hermitage and many of the other fabulous sights of that wonderful city. We spent the evening with Yaakov Frenkel, an excellent physicist whom I had met when he had come to Europe. (Five years later when I returned to Russia, this time with Ellen, my wife, he looked at her and said, "Oh, you've changed so much!")

Ruth and I spent a day or two in Moscow visiting friends and then continued on to Kharkov, where we settled in for a six-month stay. At the Physico-Technical Institute, I lectured in German, which was the scientific language of Europe at that time. This was especially true in physics, since so many of the great innovators in physics were German. I must confess that I didn't do much scientific work while I was in Kharkov. I spent most of my time talking to Lev Landau, who later became the leading theoretical physicist in the Soviet Union. He received the Nobel Prize in 1966 for his work on theories of condensed matter, particularly the strange properties of liquid helium.

Landau was an outstanding scientist who contributed to practically all fields of physics. He coauthored (with Evgeny Lifshitz) a series of books on theoretical physics that are still widely used because of their original approaches. Many people considered theirs the perfect collaboration, and when the books were published, it was said that no word in these books was written by Landau and no idea conceived by Lifshitz. Landau was a well-read person interested in every subject. He used to say, "Say something; I want to contest it." Time spent with "Dau," as he was called, was never dull.

Landau was attracted by the principles of communism, but he wanted a rational kind of Marxism and was extremely critical of the government. All the scientists at the institute were obliged to attend a weekly indoctrination lecture to hear a petty official spout the party line. Landau always sat next to me, making snide remarks and jokes as the man spoke. I was terrified we'd be arrested, but Landau wasn't frightened at all. And nothing ever happened during the time I was there. Later, during the Stalin purges, Landau spent almost a year in prison and was freed only after numerous interventions by Niels Bohr and others.

My friend Alex Weissberg was an enthusiastic, extremely boisterous fellow. He was also very knowledgeable and well educated. He could recite pages of poetry and prose by heart and, unlike me, knew all the dates of history as well as what had happened on them. On a visit to my family in Altaussee, he had impressed my mother with this knowledge of literature and art. In the Soviet Union, he was true to his word: Ruth and I stayed in a pleasant room at the guest house of the institute and got free meals. I didn't receive a salary, but there was nothing to buy in Russia in those days anyhow. I had brought enough clothes and books for the six months that we were going to be there.

One day I came back to our room to find that it had been completely ransacked. Everything we owned was gone. Weissberg was away on business, so I called the police myself. They listened to my story but took no action. We resigned ourselves to the harsh reality: our things were lost for good. When Weissberg got back a week later, he greeted me with, "Well, how do you like our country of workers and peasants now that you've been here for a while?" I said, "Not much. They've stolen everything we own."

Weissberg didn't seem at all upset. He assured me that I would have everything back within forty-eight hours. I explained that I had been to the police and that I didn't think they were going to do any more than they already had—namely, nothing. He bet me I'd have my things back in two days. All he needed was a list of the missing items. Forty-eight hours later everything we had lost was neatly laid out in our room. Every article of clothing had been cleaned, and there was even an overcoat I had forgotten to put on the list. With pleasure I paid Weissberg the money I had lost on our bet. His explanation of how he had worked this miracle was a lesson in the way things worked in the Soviet Union.

Apparently, many gangs of thieves existed all over the country, and the government knew who they were. The officials didn't mind their petty thievery, but didn't want them to become politically active, so undercover informants kept the gangs under surveillance. Weissberg had told the police that I was an important foreign scientist and that the reputation of the country and the whole communist system would be at stake unless I got my things back. The police contacted the informant for that part of town. He knew precisely where our things had been sold, so it was a simple matter to buy them back and deliver them to Weissberg.

We were visiting the Soviet Union at a particularly difficult time for the government. The experiment in the collectivization of agriculture was not working as planned. The peasants were supposed to work in large communes, or kolkhozes, instead of in small groups under a leading peasant, or kulak. But the peasants weren't educated enough to understand what was expected of them, and the government functionaries weren't able to explain the new system to them. Food was scarce, even in the cities, where there was only enough for one meal a day for most people. We could see starving peasants in the city, desperately searching for something to eat. There were reports of people dropping dead of starvation in the streets. In their zeal, the Bolsheviks had destroyed the old kulak economy; and without the kulaks, who were a kind of managerial class, no one knew how to run the farms. Agriculture had simply come to a halt.

55

Collectivization had been more successful among the communities of Germans who had settled in Russia a century earlier. They were better educated and were able to realize the kolkhoz idea. Since we shared these people's language, we were eager to see one of their communes and were pleased when we learned that a visit had been scheduled for a kolkhoz whose manager was an Austrian from Vienna. Our group consisted of Ruth, Eva Striker, an Austrian fellow named Hans Motz, and me. Eva had moved to Kharkov from Berlin when it became clear that the Nazis were soon going to take over. She had managed to get a visa to the Soviet Union by telling the authorities that she was going to marry Alex Weissberg. (Eventually they did marry, but were later divorced.)

The kolkhoz we had been invited to was in Vysoko Polie (High Field). We went there by train, sleeping in the second-class cars. We soon learned why they were called "hard cars": the seats were simple wooden boards. I put my knapsack under my head as a pillow and made sure that it was firmly placed, since my experience had taught me that nothing was secure in the USSR. I must have slept very soundly, because in the morning my head was resting on the board and my knapsack was gone, lifted during the night. Once again I had lost all my things, now also including an expensive German camera. And this time Weissberg was not there to help me get my things back. Luckily, my friends did not lose their belongings, and they were nice enough to divide what they had. We were able to put together the bare essentials for our trip.

When we arrived at Vysoko Polie, we were received like honored guests, welcomed as if we were heroes. We were introduced as proletarian comrades from Vienna who wanted to study the socialist experiment firsthand. Before we had a chance to get settled, we learned that our hosts had scheduled various tours. We were taken to the wheat and potato fields, we saw the bakery, we inspected the kitchen, and we wandered around in the living quarters. When we returned to the meeting hall, the manager appeared and said, "Now you must criticize what you have seen. Tell us what is wrong so that we can learn from our comrades from Vienna." Entirely unprepared for this, our little group had no idea what to say. Eva, who was always up to any situation, thought for a moment and then said,

"I noticed in the kitchen that the people peeling potatoes do a terrible job. So much of the potato is cut off with the skin that a lot is wasted. I noticed that often there's only a tiny piece of potato left when they are through." The manager seemed to be delighted with her criticism and thanked us profusely, assuring us that this matter would be taken care of immediately.

We were to spend the night. The accommodations consisted of one fairly large room. Our long journey in the hard car had tired us, and we slept well. The next morning was a Sunday, and when we awoke and looked out the window, we saw a church across the road. The Soviets' strong antireligious feeling may have been enunciated clearly in all the propaganda, but it had proven impossible to prevent people from attending the church services that had been such an important aspect of their lives, especially among the older generation. The bells were ringing, and we could see people entering the church. Hans Motz and Eva wanted to attend the services. I thought it was a very bad idea, but strong-minded Eva was determined to go. When she and Hans came back, they reported that the service had been crowded and that, surprisingly, the congregation had included many young people.

Later that morning the kolkhoz manager summoned me to his office. I sensed that this was an extraordinary request and had an awful feeling that something was wrong. When I entered the room, he was standing by his desk with a grave look on his face. He told me immediately that we had done something to displease the entire community. "It has been reported," he said, "that two of your party went to church. How could you do that? We are in the midst of an antireligious campaign all over the country. How does it look when our greatly admired guests from Vienna go to church here? This is a very negative thing you have done. In our view of things this is very serious."

Thinking quickly, I said, "It isn't what it appears to be. You see, we share your feelings about religion, and we were just trying to see how effective your campaign against church attendance was." He seemed to accept this explanation, and after a moment he said, "I'd like to ask you a favor, just to smooth this over. Tonight there's to be a party in your honor. I'd like you to speak at the party and counter the impression that your friends left us with this morning."

I went back to our room relieved, but so angry at Eva and Hans that I yelled at them. That evening the entire commune assembled for a gala celebration with dancing and lavish food. The manager gave a long and flowery speech and then introduced me. I got to my feet—somewhat reluctantly, I admit—and said the usual polite things about how wonderful it was to have been their guests, how much we had learned, and how much we admired the way they did things. Then, taking my cue from the previous day's success with criticism, I said that we had some reservations about the way their antireligious campaign had been proceeding. When we saw the large number of people going to church that morning, two of us decided to monitor the situation. We were alarmed at the size of the congregation and hoped that they would do better for the sake of the future of the socialist state.

At that time, as a nonbeliever, I certainly had no sympathy for the Lutheran church, which was the dominant religion in that German commune, so it was not hard to deliver this little speech. However, I must have gotten carried away by my own words, and so I exaggerated a little. It was a dramatic solution to a difficult situation. Many years later, at a party in Los Alamos when we all were asked to recall the most awkward moment of our lives, I chose that evening in the commune without hesitation.

The manager, who seemed to be relieved by my criticism, embraced me, and the party proceeded with exhibitions of folk dancing, lots of food—including a roast suckling pig—and expressions of comradeship all around. Later that night the entire village escorted us to the station to the accompaniment of a marching band.

We took several other trips during our stay in the Soviet Union. We frequently went to Moscow and Leningrad to see other scientists and friends who lived there. We traveled to Odessa to visit Guido Beck, an Austrian physicist. In Moscow I had a friend named Sascha Rumer, a Russian I had met in Göttingen. His wife, Myla, was a dancer. I will come back to him later.

Our time in Russia was extremely interesting, but I never felt comfortable there. As a matter of fact, I was always a little scared. My uneasiness did not leave me until I crossed the frontier to Germany.

⊠

# FOUR

# A Worldview and a Companion for Life from Denmark

WHEN I RETURNED to Germany, it was time to take up my Rockefeller grant. According to the terms of the grant, I was free to choose the place of study, and I decided to go to Copenhagen to work with Niels Bohr.

Ruth and I had ended our relationship while we were still in Russia, and in the aftermath of that parting I had promised myself that I was through with complicated love affairs. Physics was going to engage me completely from then on. Armed with this resolution and eager to go, I nevertheless managed to miss the train. In September 1932 I finally arrived in Copenhagen, a little late, but determined to begin my new life in science.

There on the platform was my friend Max Delbrück, like me a scholarship recipient who was going to devote the coming year to studying with Niels Bohr. Max welcomed me enthusiastically. "You are going to love it here, Viki," he told me even before I set my suitcase down. "The girls are all so good-looking. Actually, they are really beautiful—every one of them. And you are in luck because I have organized a party for tomorrow night. We are going dancing with three beauties." Before he got all the words out, I told him I was through with all that. I said I planned to concentrate on my studies, undistracted by girls or love. I was in Copenhagen to devote myself to the new physics, to learn from Bohr, to work hard. Max looked crestfallen and explained that there was no way to cancel his plans for us at this late hour. I saw immediately that he was right— it wouldn't be polite. I gave in. "Just this once," I told him solemnly.

The next evening I found myself at a large dance hall called the National Scala. There were three very pretty young women waiting for us. I don't remember the others, but the one named Ellen Tvede caught my attention. She was a professional dancer who had spent some time as a student in Dresden while I was with Heisenberg in Leipzig only a few miles away. Ellen had become part of the circle of young people who gathered around Bohr because her aunt taught Danish to most of the Bohr disciples who came from other countries to work with him. It was clear from the first that we were interested in one another. I quickly forgot my resolution about all work, no play. After that evening Ellen and I saw each other constantly, and two years later we married. We were together for fifty-seven years. Words are insufficient to describe what this companionship meant for our lives.

Max had declared that there was only one place to stay—the Pension Have, where most of the young scientists who were studying in Copenhagen lived. None of the other places were as pleasant, according to Max, and I would certainly not like the people living anywhere but at the Pension Have. Indeed, Max himself lived there. The Pension Have was exactly as he described it, a lively—perhaps a bit too lively—place where people sat up almost every night talking until two or three in the morning. I met many new friends there. One of them, George Gamow, a well-known Russian physicist who later fled the Soviet Union and spent the rest of his life in America, was famous for his extravagant sense of humor. At first I enjoyed the stimulating atmosphere at the Have. But when I eventually left, it was because the long evening discussions became too much for me, and the endless traffic noise (the Have was located at a major intersection) was too distracting.

I had come to Copenhagen to work with Niels Bohr as an apprentice in a postdoctoral program. I met him soon after I arrived in Copenhagen, and to this day he remains one of the greatest personalities I have ever known. He has influenced my life enormously, and from the beginning he made the most profound impression on me. He was my intellectual father.

Bohr was born in 1885, a little more than a century ago—and

what a century it has been. It can truly be said that these last hundred years were the best of times and the worst of times. For science the early twentieth century was an unusually creative period. Physics was revolutionized by Einstein's relativity theories and by the insights into the fundamental structure of matter that grew out of quantum mechanics. Biochemistry and molecular biology provided a deeper understanding of the origins of life, heredity, and evolution. Astronomers found it possible for the first time in history to explain the evolution of the stars and to understand how the different elements of matter were created.

Expressionism and the abstract movements in painting and sculpture showed us new ways of seeing the world around us that were in their own way as revolutionary as the new insights of the scientists. An awareness of society's responsibilities to the poor and exploited resulted in many innovative health and social programs and prepared the way for new freedom for women and an acknowledgment of children's rights. However, this was also the century in which two catastrophic world wars introduced unlimited violence and mass murder as weapons of war. Poison gas and aerial bombing of civilians were introduced in the First World War. The Second World War resulted in the development of the most devastating of all weapons, the atomic bomb, through the new scientific discoveries that became the focus of my professional life in Copenhagen.

When Bohr and I first met, all of this was several years in the future, but even then he was deeply aware of the contrast between the creative fervor in science and art and the dangers and excesses that threatened the world as we knew it. Far from being an ivory tower scientist, Bohr believed that the positive spirit of the scientific endeavor—dedication to the search for basic truth—would help mitigate the negative aspects of technical progress. It was a subject we talked about often.

Bohr grew up in a highly intellectual milieu. His father, Christian, was a professor of physiology. His mother, Ellen, came from a progressive, wealthy Jewish family. Bohr and his brother, Harald, who became a well-known mathematician, were deeply influenced by the spirit of their highly cultured family.

Niels Bohr's scientific life began in 1905, the year that Einstein published three fundamental papers: one on his special theory of relativity, one on the existence of light quanta, and one giving a proof for the existence of atoms. These works marked a turning point in physics. Bohr had the great luck to be of just the right age to enter physics at that moment. Or, more to the point, humankind was lucky that he was there at that important juncture.

He was a tall, rather heavily built man whose strong body, over-sized hands, and large skull made him look like the captain of a fishing fleet rather than a scientist. He had bushy eyebrows and a sharp, straight nose set in a long, broad face. His large head almost caused his death in 1943, when after fleeing from the Nazis he was transported from Sweden to London in a British warplane that flew very high to avoid antiaircraft fire. The helmet containing an intercom between him and the pilot was too small to fit his head, so he did not hear a warning to put on his oxygen mask. Luckily, the plane descended to a safer altitude so quickly that Bohr was not seriously affected.

When he came to an interesting point in a discussion, Bohr would often break into a captivating smile. He was never without either a cigarette or a pipe, but his smoking was always interrupted when he got involved in what was being said. During the course of a meeting he often paused to light his pipe. It became a kind of game to watch his frequent attempts to relight the cigarette or pipe that had gone out while he made a point. He was not always successful, because as he was putting the match to the pipe's bowl, a new thought would occur to him. He must have used countless matches in the course of his long lifetime.

Bohr was interested in everything, and his discourses ranged over a wide array of subjects in addition to physics. Politics was at the center of his attention at the time. He understood the threat of Nazism and the danger it posed not only to Germany and its refugees but to Denmark and the rest of the world. Indeed, any problem—large or small, scientific, political, or social—fascinated Bohr. He was interested in any phenomenon in nature. Once he explained to us how the eels swim out into the ocean from their breeding grounds deep in the Norwegian fjords. Bohr wondered how they managed to navigate around the complicated coastlines to find their way out to sea and back again to the same breeding place. Probably, he said, they didn't think about it, just as the centipede would be paralyzed if it thought for even a moment about which foot to move next.

Bohr loved paradoxes in nature and of the mind. What attracted him most were unsolved problems and the many seeming contradictions that we observe in nature. The more tantalizing the problem, the more he was drawn to it. When faced with a deep riddle, an apparently insoluble problem, he always said, "Every great and deep difficulty bears in itself its own solution. It forces us to change our thinking in order to find it." Quantum mechanics, he loved to say, was the best example of this principle. The quantum, the wave and particle nature of the electron and other entities, did not make sense to those who used classical methods of thinking. Only when our approach was thoroughly changed did the answers emerge.

Bohr participated in the discussions among his disciples with great intensity and attention. As a result, he often helped us clarify our thinking. "No," he often said after a presentation, "you must explain that again." And most of the time he wasn't satisfied until we had reformulated our explanations several times in several ways. In order not to discourage us when he felt we were heading in the wrong direction, he used to say in his peculiar "Bohrish" manner, "It is not to criticize, only to learn what you meant."

Bohr made important contributions to the new method of approaching physics that ultimately led to quantum mechanics. His 1913 papers on atomic structure were a turning point in the history of atomic physics. He explained how our approach to the dynamics of the atom must differ from the old view if we are to explain the atom's stability, its radiation, and its characteristic properties. In these papers he showed that clever applications of some of the new ideas of Planck and Einstein regarding the quantum of energy would determine the size and general structure of the hydrogen atom and explain quantitatively the light emitted and absorbed by it.

Strangely enough, Bohr's papers did not get much attention in the physics community. Those few who studied his work were divided in their response. Some were enthusiastic and thought the problem of atomic structure had been solved. But, in fact, these papers did not solve the problem from a fundamental point of view. Instead, they indicated that the explanation of atomic properties necessitated a radically new way of thinking about nature.

It was not until the mid-1920s that this new way of looking at atomic phenomena received its logical formulation through the work of Heisenberg, Schrödinger, Born, Pauli, and Dirac. This was accomplished with the constant prodding and help of Niels Bohr. But Bohr considered his 1913 papers only a tentative approach that left many questions open. In a letter written to his brother in 1912, he said, "It could be that what I found out in just a small glimpse of the structure of atoms may perhaps be a little bit of the true reality."

Other physicists were very skeptical. There is a story, which may be apocryphal, that the distinguished scientists Max von Laue and Otto Stern discussed Bohr's paper on the hydrogen atom during a promenade on the Ütliberg near Zurich. They swore they would give up physics if these newfangled ideas turned out to be relevant. They called their decision the "Ütli oath," alluding to the famous Rütli oath in 1291 A.D. when the first Swiss cantons joined in a Swiss confederacy and swore eternal trust and independence. In contrast to the faithful Swiss, who remain true to their oath after nearly 700 years, von Laue and Stern remained active physicists even after Bohr's theories proved deeply relevant.

In the 1920s Bohr made his essential contributions to quantum mechanics. He received funds from the Rockefeller Foundation and from Danish sources to build the Institute for Theoretical Physics, or the Bohr Institute. He had enough money to invite a select group of young theoretical physicists from many different countries for varying tenures. Among them were Oskar Klein from Sweden; Hendrik Kramers and Hendrik Casimir from Holland; Werner Heisenberg, Pascual Jordan, and Carl Friedrich von Weizsacker from Germany; Felix Bloch from Switzerland; Wolfgang Pauli from Austria; P. A. M. Dirac and Nevill Mott from England; John Slater from America; and Lev Landau and George Gamow from Russia. The only woman in the group was Lise Meitner, an experimentalist who visited occasionally. This truly international crowd assembled at the Bohr Institute was, perhaps, the first example of the kind of international scientific effort that Bohr championed all his life.

The years from 1924 to 1930 were a heroic period for physics, encompassing the birth of quantum mechanics. During those years Bohr produced only a few papers that bear his name alone. Instead, he developed a new way of working collaboratively. The foundations of quantum mechanics were conceived, discussed, and then written down mostly in papers by his collaborators, but the ideas were, to a great extent, stimulated by him. Werner Heisenberg's formulation of the fundamental limits of classical concepts in his uncertainty principle is an example of Niels Bohr's collaboration with his younger colleagues. Over an extended period of time, Bohr and Heisenberg held numerous discussions on this subject at the Bohr Institute, in correspondence, and in conversations during excursions along the Danish Coast or in the German Alps. Bohr, in his Socratic way, always asked the relevant questions and pointed to the depths of the problems and the plenitude of the phenomena. Then in the minds of the best of his disciples, the new ideas grew and took shape. Obviously, such a relationship also resulted in a sharing of philosophy; common outlook; wide-ranging human, artistic, and literary concerns; and ultimately, in many cases, in deep personal friendships.

Imagine the atmosphere, the intellectual ferment that pervaded the institute during those years. Bohr was at the height of his powers. The style he created was called *Der Kopenhagener Geist* (the spirit of Copenhagen). Among his collaborators he was admired extravagantly, and he was always active—talking, creating, living, and working as an equal in a group of young, optimistic, jocular, enthusiastic people who were involved in approaching the deepest riddles of nature with a spirit of attack, freed from conventional bonds in a creative climate that can hardly be described.

In this period Bohr and his colleagues penetrated the secrets of the inner workings of nature. In the course of a few years they laid the foundations for a science dealing with atomic phenomena that grew into the vast body of knowledge available today. A variation of Churchill's famous statement about the Royal Air Force to the House of Commons in 1940 is appropriate here: Never have so few done so much for so many in so little time. During this period Bohr also developed his concept of complementarity. It originated from the discovery that electrons (or any other particles) behave sometimes as waves and sometimes as particles—seemingly contradictory properties. An electron is neither a wave nor a particle, but it exhibits one or the other set of properties under certain well-defined conditions. The systematics of these dual roles represent the essence of quantum mechanics. Bohr used the term *complementarity* for the apparent contradiction between the two mutually exclusive properties. They are complementary aspects of the same object, the electron. Both are necessary to encompass the totality of its properties. David Bohm has expressed the situation as follows: "The electron may be regarded as an entity that has potentialities for developing either particle properties or wave properties, depending on the type of instruments with which it interacts."

Niels Bohr taught me to appreciate the importance of different, even seemingly contradictory avenues to human experience. He generalized the idea of complementarity to fields of human experiences outside of physics, such as ethics, music, art, and religion. His ideas had an enormous impact on my own attitude toward human concerns. As examples of complementarity, he often mentioned the scientific versus the poetic approach to an experience or justice versus compassion or psychology versus neurophysiology. The pair aspect is by no means essential—on the contrary, there are often many different avenues for dealing with such problems.

Bohr used to point to a cubist painting in his house, in which a human being is depicted in many self-contradictory ways that show the different aspects of the person simultaneously. Then he explained that to grasp the object fully one needs all these different perspectives, although they do not seem to make sense from a superficial point of view.

Bohr was an extraordinary person, but his immediate family was also exceptional. His wife, Margrethe, was an impressive woman of queenly appearance and beauty that did not leave her even in old age. She did not understand physics, but she was deeply aware of the importance and depth of Bohr's work. She created a warm human environment in which Bohr could be as creative as he wanted to be. They had five sons. The first died in his teens in a sailing accident, but the rest grew up to be accomplished people. Aage Bohr became a physicist and won the Nobel Prize, like his father. (It is rare for the children of great thinkers to achieve greatness in their own right.) I suspect that Niels's deep understanding of human nature and Margrethe's sensitive support encouraged their children to reach their full potential and protected them from the psychological handicaps of being the offspring of a powerful personality.

Bohr became a father figure to all of us who worked with him. He had established the habit of inviting each of his new apprentices to go for a walk with him shortly after arriving at the institute. On these excursions he would try to draw the young scientist out to find out how he felt about life and what he knew about science. Although we were undoubtedly very much in awe of him, he managed to get us to talk openly about ourselves. Early in my walk with him he asked me, quite naturally, "Well, how do you like it here?" With the self-righteousness and arrogance of the young, I said I thought it was wonderful and I liked the people I had met, but that I was shocked by the amount of fooling around, the jokes and pranks that were so much a part of everyday life at the institute. "I expect scientists to be more serious," I declared. Bohr responded in a way that I later realized was typical of him: "There are some things that are so serious you can only joke about them."

In spite of this one close encounter that each of us experienced, getting to know Bohr took some time. At first we apprentices saw him mainly at a slight remove, but later the contacts became closer. Every week discussions were held in the small lecture hall at the institute. Bohr always sat in the same seat in the right-hand corner of the first row, and we sat in rows all around him and talked about the things we had been working on.

I came to the Bohr Institute when quantum mechanics was already an established theory. I missed those exciting early years when every Ph.D. thesis opened up a new field of application for quantum mechanics. When I joined the institute, new problems were being discussed, such as the extension of quantum mechanics to the electromagnetic field and the study of the inner structure of the atomic nucleus. Nevertheless, Bohr always came back to the foundations of quantum mechanics and loved to explain and discuss it with members or guests of the institute.

\*  \*  \*

The group I met at the institute contained the usual mix of people from many countries. Among our group of about fifteen there were a few who made up what we called "the inner circle," among whom were George Placzek, Felix Bloch, George Gamow, E. J. Williams, Subramanian Chandrasekhar, and Homi Bhabha.

Chandrasekhar, known as "Chandra," later became a famous astrophysicist and won the Nobel Prize. He explained many important features of the development of stars and is best known for the Chandrasekhar limit, which describes the maximum mass of a star, beyond which the star ends its life by exploding into a supernova and collapsing into a black hole. Chandra was a quiet, somewhat shy man with a beautiful face and wonderful eyes. Placzek referred to him as "the boy with the doe eyes."

Homi Bhabha became a leading figure in Indian science and founded and directed a large research institute in Bombay. He was also thoroughly versed in Western culture and was a painter of some renown. We often listened to music together. Once he pointed out some special points in the late Beethoven quartets that I, who had been raised with this music, had not appreciated until he called my attention to them.

George Placzek, a native of Czechoslovakia who grew up in Vienna, was one of the most extraordinary people I have ever known. We considered him a wise man with whom we were eager to discuss personal, political, and scientific problems. Placzek had an especially clear insight into the problems of the time, whether in physics or politics. But for some reason he was never able to find the right style of life for himself, and his personal life was not a happy one.

Impeccably honest and loyal to his friends, Placzek was known for his intellect and scientific prowess. His somewhat outrageous sense of humor and lifestyle inspired dozens of "Placzek stories." He was a truly unique man, the kind of person who is constantly involved in some sort of antic adventure. For example, after being in Copenhagen for some time, Placzek got a little apartment in the institute. He used to sleep very late, often until noon, when he would get up and go down to get his mail. The sight of Placzek still in his rumpled pajamas at midday in the library offended several of the female students, who complained to Mrs. Bohr. She asked me, as

Placzek's friend, to intervene. Placzek assured me that there was never anyone there when he came for his mail. All the other physicists arrived later, he insisted. I said that there were always several students in the library where the mail was kept. His eyes were wide with innocence as he said, "Oh, you mean Danes."

As a devout bachelor, Placzek was horribly afraid of the famous, unwritten Bohr Institute Rule: Any physicist working with Bohr was certain to be married after no more than two years. When his second year was nearly over, Placzek began to worry that his happy days as a single man were about to end. Since I was engaged to Ellen after only a year in Copenhagen, I decided to give him the year that was still owed me by tradition. In a solemn ceremony I transferred it to him so that he could be secure for one year longer. Much later when he did wed, the marriage was an unhappy one, so perhaps his instincts in this regard were sound.

Without being assigned this task, the inner circle at the institute assumed the responsibility of protecting Bohr from visitors we considered unworthy of coming into contact with him. I'm afraid that in our proprietary zeal we were sometimes rather severe in this regard and may have discouraged some good physicists from working with Bohr.

When we were not engaged in science, we often went out with Bohr, especially to the movies, which he loved. He had a favorite actress, Annie Ondra, a charming starlet. Bohr insisted on going to every one of her movies, and we tagged along, happy to be invited. He also liked westerns, but he was sometimes unable to suspend his scientific sense of reality. I remember once, after we had seen a cowboy movie, he said, "I don't think the action was really plausible. The hero just happens to arrive when the girl is in dire need. Well, I suppose that could happen. But then the hero is able to kill all the people who molested her. Well, that is also possible perhaps. And it is possible that he takes her on his horse and gallops through the desert with her. I can even believe that the horse would be able to carry them both and manage to jump the river." "So, what's the trouble?" we asked. "What don't you believe?" "What I can't believe is that during all of this, a photographer would just happen to be there." And so we learned something about Bohr's sense of humor.

Bohr took us to many other popular places, including the amusement park in the Tivoli Gardens, where he displayed an incredible talent for discovering how the magicians did their tricks.

Back at the laboratory Bohr was invariably at work on a paper. He used his apprentices mercilessly to help him write his papers, selecting one "unlucky" person whom we named "The Victim." When I first arrived in Copenhagen, the chosen victim was Leon Rosenfeld. I had only been at the institute a short time when Rosenfeld was called home because of a family emergency. As I was the newest member of the inner circle, I was chosen to take Rosenfeld's place.

Bohr was working on an interesting problem at that moment. It was a period during which quantum mechanics was being applied not only to particles such as electrons but also to the electromagnetic field. It therefore seemed necessary to Bohr to find similar physical arguments to explain why there are limits to the simultaneous determination of certain electromagnetic magnitudes. The velocity and the position of a particle cannot be determined simultaneously. The reasons for this were given by Heisenberg, who showed that any measurement of the position of a particle necessarily influences its velocity and vice versa. Bohr could show that a similar mutual exclusion exists in, for example, the simultaneous measurement of the electric field in a given direction and the magnetic field in the perpendicular direction within a small volume.

As the temporary victim, I had only a limited involvement in these rather intricate investigations. Bohr's main work was done with Leon Rosenfeld, but my contribution was generously acknowledged in the paper Bohr published. The victim's life was not his own; he was at the mercy of Bohr's whims. When Bohr decided that he wanted to work on a paper, the victim was called into his office and seated at a desk. There he stayed throughout the day while Bohr roamed around the room circling the victim's desk every few minutes. This gave me particular trouble because I have never been able to sit still for any length of time. By then I had already developed my lifelong habit of pacing while I worked. But Bohr wouldn't have it. "Only one of us is allowed to move" was his rule, so I sat there day after day in agony.

Bohr's scientific method consisted of thinking aloud. It was the victim's job to speak up when he couldn't understand what Bohr was saying or didn't think it was clearly formulated. This, in itself, took a certain amount of courage. Then Bohr would rework the formulation, and finally there came the moment when it was necessary to write something down. Bohr would begin dictating an extremely long sentence in German. Usually I would try to get him to make it into two sentences, but Bohr would insist—and naturally, he rarely lost. Often when he started to formulate a sentence, he would interrupt himself to begin a better formulation. Dirac's career as victim ended abruptly after half an hour because he said to Bohr, "I wonder whether you were ever told in grammar school never to write a sentence before thinking it through completely." When the victim got over the feeling of being used, he realized that he had been given a glorious opportunity. He was witness to the mind of Niels Bohr in action. It was a chance to learn how Bohr thought and how he constructed the solutions to the problems he posed. In retrospect I am grateful for having had the privilege of being "victimized" by Bohr.

One of the perks of the victim's position was lunching every day with Bohr and Margrethe. We would talk about politics and other matters because Mrs. Bohr was not able to discuss physics. But Bohr depended on his wife and used her as a sounding board for matters of politics or human problems. I can still hear the way he called, "Margrethe, Margrethe, come hear what Weisskopf [or Bloch or Rosenfeld] says." The Bohr inner circle often imitated this and used it as an inside joke. I frequently used it to get Ellen's attention.

When I first arrived in Copenhagen, the Bohrs lived in a house on the grounds of the Institute of Theoretical Physics. But a short time later, the Danish philosopher Harald Höffding died. As Denmark's most distinguished scholar, he had occupied the mansion that had been given to the state by the Carlsberg Beer Company family after the death of their founder. When Höffding died, the family, longtime supporters not only of art but also of science, designated Bohr as the next inhabitant of the Schloss Carlsberg, as we called it, although its real name was Aeres Boligen (Honor Mansion).

From my lodging it took nearly half an hour by streetcar to get to the mansion. One evening at six o'clock, my usual quitting time, Bohr and I were still deep in discussion. I had an appointment that night and had to leave promptly, so Bohr walked me to the streetcar stop, about five minutes from his house. We walked and he talked. When we got there, the streetcar was approaching. It stopped and I climbed onto the steps. But Bohr was not finished. Oblivious to the people sitting in the car, he went right on with what he had been saying while I stood on the steps. Everyone knew who Bohr was, even the motorman, who made no move to start the car. He was listening with what seemed like rapt attention while Bohr talked for several minutes about certain subtle properties of the electron. Finally Bohr was through and the streetcar started. I walked to my seat under the eyes of the passengers, who looked at me as if I were a messenger from a special world, a person chosen to work with the great Niels Bohr.

People used to say that Bohr was the uncrowned king of Denmark. He was about forty-eight years old when I worked with him, at the peak of his intellectual powers and involved in many aspects of Danish life.

The Rockefeller grant that supported me in Copenhagen was originally meant for half a year in Copenhagen with Bohr and half a year in Cambridge, England with Dirac. But I loved Copenhagen so much—both because of the inspiring environment at Bohr's institute and because I had met Ellen there—that I ended up staying more than six months. When I left for Cambridge in late spring, it was for a much shorter time than I had planned.

I knew a number of people in Cambridge, notably Rudolf Peierls, a refugee from Hitler's Germany. He and his wife had spent their Rockefeller-grant year entirely in England. I also made the acquaintance of one of the great British experimental physicists, Patrick Blackett, and his wife, Pat. He played an important role later as a scientist in the Second World War helping the government in weapons policy. His political views were rather left of center, and I found him to be a kindred spirit. When I arrived, I rented a nice room and soon discovered that Cambridge was a wonderful city in which to live and work. I learned a lot of physics, but not the way I had intended. I had planned to work with Dirac, but he was a rather withdrawn person who didn't talk much and was not much interested in the work of other scientists.

Fortunately, I was able to take advantage of my acquaintanceship with Peierls, who was a few years older than I and from whom I learned a great deal. Two or three years can make a big difference when one is young. Peierls introduced me to relativistic field theory and he taught me how to make calculations with the Dirac equation, a skill referred to then as "alpha gymnastics." Peierls's stipend was $200 a month, whereas mine, to my slight annoyance, was $150. I hasten to add that either amount represented, at that time, undreamed-of riches for an average European. The reason Peierls was entitled to more money was that he was married, whereas I was still only engaged. The Rockefeller Foundation asked us to send reports of our achievements during the time of support. As proof of some of his activities, Peierls sent the foundation an announcement of the birth of his first child. We subsequently heard that the Rockefeller trustees in New York didn't appreciate Peierls's sense of humor.

My stay in Cambridge gave me some insights into British life and the way British universities were run. In contrast to the central European universities, there was much less emphasis on lectures to a large audience of students and more direct contact with tutors. While in Cambridge I also met two of the greatest experimental physicists of our time. One was Ernest Rutherford, who discovered the atomic nucleus in 1911 and suggested that the atom is similar to a planetary system, with electrons circling the nucleus. The other was the Russian scientist, Peter Kapitza. He had been given a lab in Cambridge where he studied the behavior of matter in very strong magnetic fields and at low temperature.

In the spring of 1933 something took place that decisively influenced my further development as a scientist: a letter came from Wolfgang Pauli asking me to be his assistant in Zurich. I was to replace Hendrik Casimir, whom Paul Ehrenfest had called back to Leiden. What could be better for a young physicist than to work with Pauli, one of the pioneers in the development of quantum mechanics? It was the fulfillment of a dream and one of the greatest things that happened to me during my entire professional life. For a brief moment I wondered why he had picked me and not some other more experienced scientist like Hans Bethe. But I put this out of my mind. Later I found out why I had been chosen.

I immediately showed the letter to Peierls, who had been Pauli's assistant earlier. I had learned a lot of science from Peierls; now I was asking him for help in psychological matters. Pauli had the reputation of being a difficult master, and I thought Peierls could advise me on how to handle him. Peierls told me not to worry. "Pauli is actually a very nice, almost childish person," he said. "His often-mentioned disagreeable characteristics come more from his honesty than from any malevolence. He always says what he thinks, and that's what sometimes offends people. But it can be very easy to work with a man who always lets you know what he thinks."

However, he added, "Be very careful when you give a talk at a seminar which Pauli is planning to attend. He has a habit of interrupting the speaker repeatedly, saying, 'This is all wrong' or something of that sort. This isn't hostility. He simply can't hold back what he's thinking. It just comes out of his mouth." Peierls suggested that I should go to Pauli the morning of the lecture and show him what I planned to say. "If he doesn't like it, he'll tell you in the strongest possible terms," Peierls said. "Then in the afternoon you will say exactly what you planned to say in the first place. You won't need to change a word unless he has really convinced you. Pauli will sit in the first row, and when you come to the critical points, he will almost inaudibly mumble to himself, 'I've told him already, I've told him anyway,' and so it won't be so bad."

I was ready for Pauli.

FIVE

⊠

# Adventures with Pauli

WHEN THE TERM in Cambridge ended in late summer, I went to Austria for a vacation. Then in the fall I moved to Zurich, which I had always considered a wonderful city with many cultural amenities as well as extraordinary opportunities for skiing and hiking in the surrounding Alps. I was terribly excited about being able to live in such a marvelous place while having the incomparable experience of being the assistant to one of the world's most important physicists.

Ellen and I had decided to separate for a year to test our resolve to marry, so I moved to Zurich alone. I went there still wondering why I had been chosen instead of some other, more experienced young scientist. I soon found out. When I arrived at the institute, I knocked at the door of Pauli's office until I heard a faint voice saying, "Come in." There at the far end of the room I saw Pauli sitting at his desk. "Wait, wait," he said. "I have to finish this calculation. (Erst muss ich fertig ixen)." So I waited a few minutes. Finally, he lifted his head and said, "Who are you?" I answered, "I am Weisskopf. You asked me to be your assistant." He replied, "Oh, yes. I really wanted Bethe, but he works on solid state theory, which I don't like, although I started it." I finally had the answer to the nagging question I had carried with me to Zurich.

I made a contract with Pauli. I said, "Of course, I am more than delighted to work for you. But please, that new stuff you are working on, the Klein-Kaluza approach to general relativity—I am unable to understand that, and I just don't want to deal with it. I am ready to work on everything else." He accepted the conditions because he

75

was already somewhat bored by the work. (Today it is the dernier cri of the most sophisticated particle theorists.) Pauli gave me some problem to study—I no longer remember what it was—and after a week he asked me what I had done about it. I showed him my solution, and he said, "I should have taken Bethe after all." Since I had been well prepared by Peierls for events like this, I was more amused than taken aback. I decided to take my years with Pauli for what they offered—a challenge and a chance to work with a great man in order to get a deeper understanding of physics.

My time in Zurich was also interesting for other, more ominous reasons. These were the first years of the Hitler regime, and a stream of refugees had come to Switzerland. Zurich quickly became an important center of German culture because so many of the persecuted artists moved there. The Zurich *Schauspielhaus* in those years was the best German theater in the whole German-speaking region. The refugee actors performed for very little money, and Zurich and its citizens benefited from being able to see a new play every week put on by excellent performers under the direction of the eminent Kurt Hirschfeld, whom I had met in Berlin and who introduced me to many of the famous exiles, including Bertolt Brecht.

During my first year I lived at a pension, or boardinghouse, presided over by a Mrs. Biske. She was a Russian refugee who had come to Switzerland in 1905 after the pogroms that also brought many refugees to the United States that year. Among the other people staying at Mrs. Biske's was a man named Heinz Kurella and his girlfriend, Charlotte. He was an avowed Communist who had fled Germany and was secretly working for a weekly Communist newsletter, the *International Press Correspondence* (IMPRECOR). Because he read and then summarized articles from the newspapers of many different countries, he was aware of what was going on all over the world. I struck up a friendship with him and Charlotte, whom we all called "Red Countess" because of her aristocratic good looks. Although Heinz and I disagreed on many political issues, it was very interesting and stimulating to talk to him and Charlotte. We often played cards together in the evenings, and he meticulously entered our scores in a notebook. Later I was to regret this completely innocent habit of his.

One day the Swiss police got wind of Kurella's Communist activities. When he saw police officers coming up the steps of the pension, he managed to get out the back door and escape with the help of Mrs. Biske. But Charlotte was caught and arrested. This event placed Mrs. Biske once more in a role she had played with distinction in 1905 when her husband had been imprisoned in Russia. She had helped Kurella escape, and now she was bringing food and consolation to Charlotte, who was eventually freed and expelled from Switzerland. Kurella made his way to Prague and eventually back to Russia, where, like so many others, he was imprisoned and executed during the Stalin purges.

In 1936, when I was back in Copenhagen, a friend of Kurella's came to visit me and gave me all of his books, the ones that had gotten him in trouble with the Stalinists. They included the works of Nikolay Bukharin, Leon Trotsky, and other Bolshevik leaders. The friend told me his name was Edgar, but I knew that was a cover name. He and I talked about the Stalin purges, and I told him I was afraid that if he went back to Moscow he would be in real trouble. But he said, "I have no choice. I've been a Communist all my life, I worked for communism, and I can't see breaking with them now." I took Edgar to the station, and somehow I knew that I would never see him again.

After my first year in Zurich, Ellen and I decided to get married. I had to ask Pauli for permission to take ten days off and go to Copenhagen. In general, Pauli didn't like his assistants to leave in the middle of the term, so I was anxious as I walked to his office. When I asked for a leave, he quickly asked me the purpose of my trip. When I told him I wanted to get married, he said, to my great surprise, "Oh, I approve, because I am also getting married." And when I told him whom I was going to marry, he said, with a smile, "Oh, the Daughter of the Regiment," referring to Ellen's popularity among Bohr's disciples.

Our wedding in Copenhagen on 4 September 1934 was a merry event among our many friends there. The official ceremony at the city hall was purely formal, with only the few required witnesses. But George Placzek organized a big party for the evening, featuring a special "wedding ceremony."

Max Delbrück was the "minister." First, Ellen and I had to undergo a test to prove that we were ready to make such an important decision. We had to walk a straight line on the floor, which was not easy because we had drunk quite a lot of champagne.

The actual ceremony began with an appropriate selection from *The Threepenny Opera*: the wedding of Mack the Knife and Polly Peachum.

> *Mack:* Do you see the moon over Soho?
> *Polly:* I see it, love. Do you feel my heart beating, beloved?
> *Mack:* I feel it, beloved.
> *Polly:* Where you go, there I too want to go.
> *Mack:* And where you stay there I too wish to be.
> *Polly and Mack together:* And if we have no license from the registrar
> And no flowers upon the altar
> And though there is no myrtle in your hair
> Love will or will not endure here or in some other place.

Happily, Ellen and I could say that our love lasted undiminished for more than half a century.

Pauli's marriage was his second. His first had lasted only six months. His second wife, Franca, was a wonderful person. Pauli could not have been an easy person to live with, and Franca made his life much more pleasant. She provided him with a home where he felt at ease and could pursue his many interests in a calm atmosphere.

Ellen and I found an apartment in Zurich. Like me, she loved the city, with its many cultural attractions and the mountains, the hiking, and the skiing. We also developed many warm friendships with both the Swiss and the refugees from Nazi Germany. We lived not far from the station and went to the many excellent ski slopes that were less than an hour's train ride from Zurich. During Christmas and spring vacations we went to places in western Austria not only for the wonderful skiing but also because Nazi Germany had closed the frontiers to Austria as a retaliation against the Austrian government for their anti-Nazi measures. Since the Austrians depended mostly on German customers, their ski resorts suffered from these restrictions, and people from other countries made an effort to ski there in order to help them out.

We made one unforgettable Christmas trip to an isolated resort called Hoch-Krumbach in the Arlberg region. We traveled by train to Bregenz, a town near the Swiss border, and by bus to the village of Schröken, where we stayed overnight. The next morning a horse-drawn sleigh took us to Hoch-Krumbach. We arrived at the inn in Schröken on Christmas Eve. The village was covered by a heavy blanket of snow, and it was very cold. As we entered the well-heated main hall of the inn, we saw a huge decorated Christmas tree lit with what seemed like hundreds of real candles. Around the base of the tree were the villagers and their children praying. Then they sang "Silent Night." Every Christmas we talk about that unforgettable scene in rural Austria.

During that vacation we visited the Paulis, who were staying at a resort a few miles away. We started our return on skis a little later than we should have. A heavy snowstorm came up, and the skiing became difficult. Very quickly darkness fell, and the wind pushed us off the trail, which was hidden under the deep snow. I was worried that we wouldn't make it, spending the night outside in the cold could have proven life-threatening. Just when I was beginning to feel genuine fear, we heard voices and saw the lights of Hoch-Krumbach, and soon we were back safe at the inn.

Back in Zurich our way of living was quite different from that of the Swiss people in our apartment building. They objected, for one thing, to my piano playing. I was told that I would not be allowed to play piano after ten o'clock in the evening. Our neighbor below didn't understand why I had to play the same thing over and over and complained that it disturbed him to hear my repetition. I was practicing a complicated piece by Brahms and having a hard time getting it right, but he wouldn't listen to my explanation. Relations with our neighbors were not improved when one evening we came home late from a party and spent some time in the entry hall talking with a friend. Apparently, at one point I leaned against the panel of doorbells. It was midnight, and the good people of our building were awakened from their sleep. We heard about this scandal for some time.

At the institute Pauli had given me a very important problem to solve: I was to recalculate the internal energy of an electron—the so-called self energy—by making use of Dirac's new field theory. Because of Einstein's theory of relativity, the question of the mass of the electron was known to be connected to its internal energy by the famous formula $E = Mc^2$. Dirac's theory deals with the interactions of electrons and electromagnetic fields, making use of both quantum mechanics and the theory of relativity.

The internal energy of the electron posed an awkward problem for physicists. The difficulty is related to the electric field surrounding the electron. The smaller the radius of the electron, the stronger is the electric field in its immediate neighborhood, which leads to a higher energy. If the electron were an infinitely small point, its internal energy, and therefore its mass, would be infinitely large. In order to get the actual mass, one was forced to assume a relatively large radius, which directly contradicted many observations.

My calculation resulted in one of the darker moments of my professional life. I made one of those minor errors that can have a big effect. Somehow I confused a plus sign with a minus sign, and the self energy I came up with turned out to be very large. The application of the field theory seemed to make things even worse compared to previous calculations. This was the conclusion I published in my paper.

A few weeks after the paper appeared I received a letter from the physicist Wendell Furry, who was then working with J. Robert Oppenheimer. Furry had realized that I had made a simple mistake of a sign in my calculation. When the error was corrected, the new field theory improved things considerably. It turned out that one can obtain a fairly correct mass for the electron by altering the field only at extremely small distances, where field theory is probably no longer applicable. I felt foolish for having made such a silly mistake in a fundamental problem, and I was depressed over having published my error. I told Pauli that I wanted to give up physics, that I would never survive this blemish on my professional record. Pauli tried to console me in his usual way. "Don't take it too hard," he said. "Many people have published papers with errors in them. Of course, I never did."

I wrote to Furry and suggested that he publish his result under his name, or if he didn't want to do that, we could copublish a corrected paper. He answered that I should publish the paper under my own name, mentioning him as the person who drew my attention to the error. That is what I did, and since then the method of calculating the self energy of the electron is ascribed to me, although I have always felt that it should have been credited to both of us. Furry's gentlemanly attitude was what one expected in those days. (I remember once showing Pauli a newly published paper on a subject in which he was interested. He said, "Yes, I thought of that too, but I am glad he worked it out so that I don't need to do it myself.") Not everyone was that generous then, but there was a much more relaxed attitude about such things than there is today.

After the Second World War the competition among physicists became so strong that things changed radically. Physics had become an extremely popular, even glamorous field because of its many practical applications to war. Radar saved England, and the atomic bomb ended the war. The number of physicists grew very quickly during the postwar years. This growth was accompanied by a fierce new competition for jobs, and the gentlemanly attitude that had pervaded the field before the war became rare.

I collaborated with Pauli on some research into a new subject of great interest: antiparticles. As mentioned before, the existence of positrons, the antiparticles of electrons, was predicted by Dirac from purely theoretical arguments and found in nature by Carl Anderson only a few years later. The most astounding property of antiparticles is their interaction with the corresponding particles. If a positron hits an electron, both of them disappear and their masses are transformed into light or other forms of energy. This process is called "annihilation." The reverse process is even more surprising. Under certain conditions, pure energy—a light beam, for instance—can be transformed into a pair consisting of a particle and its antiparticle. An electron-positron pair can be created by light energy. This process is called "pair creation."

The method by which Dirac derived the existence of the positron and pair creation and annihilation was not quite satisfactory. When he formulated his famous equation describing the dynamics of an electron, he had to assume certain strange properties of the empty space, and Pauli disliked these assumptions. Today one can derive all of this in a much more satisfactory way, but this was unknown when I was working with Pauli.

The charged particles known at that time all possessed an intrinsic spin. Massive particles without a spin (mesons) are known to exist today, but they were unknown then. There was only one equation, the so-called Klein-Gordon equation that described charged particles without spin. I was interested in this equation. It was simpler than Dirac's electron equation, and I played around with it, although there seemed to be no charged particle in nature devoid of spin. I discovered that such particles, if they existed, would appear in pairs of opposite charge. I found some indications that these pairs could be created by light or annihilated into light, without making any strange assumptions about the empty space.

However, I was unable to perform the calculations and approached Pauli to help me out. He was in a very bad mood that day. I tried to explain my difficulties and tentative conclusions, but he kept impatiently saying how silly he found my remarks. In a kind of amused desperation I quoted to him a verse from Wagner's opera, *Die Meistersinger:* "Ach Meister, warum soviel Eifer und so wenig Ruh, Mich dünkt euer Urteil wäre reifer, hörtet ihr besser zu (Oh master, why so much excitement and so little repose; I believe your judgment would be more mature if you would listen better)." Pauli looked up at me and asked, "What is that?" I told him it was from *Die Meistersinger,* whereupon he replied, "Oh, Wagner I don't like at all." So ended that discussion.

The next day he was in a better mood, and I repeated my concerns about the calculations. This time he said, "That's interesting. Why didn't you tell me about this yesterday?" From then on there began a marvelous collaboration with Pauli in which he taught me how to deal with such problems. We found that the quantum electrodynamics of spinless particles indeed leads to antiparticles, pair creation, and annihilation. This all comes about without any of the assumptions Dirac had to make about the vacuum in order to get these phenomena from his equations. As Pauli had never liked the way Dirac came to his conclusions, he referred to our paper as "the anti-Dirac" paper.

Pauli also asked me to calculate the pair creation and annihilation cross sections of spinless particles according to our theory. The calculation was not too different from the one for ordinary electrons and positrons that Bethe and his collaborators had carried out in 1934. I met Bethe at that time at a conference in Copenhagen and asked him to show me how to do the calculations. I wondered how long it would take to get to the final result, and he told me, "It would take me a few days. It will take *you* a few weeks." It did. Moreover, I made a mistake; my result was four times larger than the correct one. Once again I had proven that Pauli should have taken Bethe.

Our paper was little more than an interesting formal exercise at that time, because no particles with zero spin were yet known to exist, although the Japanese physicist Hideki Yukawa had proposed such particles as carriers of the nuclear force. We published our paper in the honorable but little-read journal *Helvetica Physica Acta*. A few years later such particles, the pi-mesons, were actually discovered, and our results acquired some practical importance.

Some years after I left Pauli, he used ideas from our research for his famous proof of the connection between spin and statistics. The statistics of particles determine their behavior when several or many particles of the same kind interact and form structures. There are two types of statistics, the Fermi statistics and the Bose statistics. Under the Fermi statistics, two particles of the same kind cannot both be in the same quantum state, a feature that is often referred to as the "exclusion principle." Under the Bose statistics, particles prefer to be in the same quantum state. It was found that particles with half-integer spin always fulfill the exclusion principle, whereas particles with integer spin act according to the Bose statistics, but nobody knew why until Pauli explained the connection. He discovered the exclusion principle in 1925 from a careful analysis of atomic spectra and was not satisfied until fifteen years later, when he was able to show how it followed by necessity from quantum field theory.

The numerous Pauli anecdotes circulating among physicists give a distorted impression of Pauli's personality. They make him out to be a mean man who wanted to hurt his weaker colleagues. Nothing could be further from the truth. Pauli's occasional and highly publicized roughness was an expression of his dislike of half-truths and sloppy thinking, but it was never meant personally. Pauli possessed an almost childlike honesty, and always expressed his true thoughts directly. Although there is nothing more reassuring than to live and work with someone who says exactly what is on his mind, it does take some getting used to. I am certain, however, that he never wanted to hurt anyone, although he sometimes did so unintentionally. When Ehrenfest met Pauli for the first time, Ehrenfest said, "I like your physics better than I like you." To this Pauli responded, "With you, Professor Ehrenfest, it is just the other way around." Nevertheless, they became very close friends.

Then there is the famous "Pauli effect," which followed Pauli around wherever he went. Or so the story goes. It was said that "even his presence in the same town could make experiments go wrong." Once when Pauli was expected at a laboratory in Milan, the physicists decided to play a joke on him. They wired a door so that when Pauli opened it, a loud explosion would be heard inside the lab. They tested it several times to make sure it worked. But when Pauli arrived and opened the door, nothing happened. The Pauli effect had defused the device and protected him from the joke.

Pauli placed little value on pedagogical efforts. Once when I told him I wanted to write a popular article on the theory of the positron, he said, "Yes, but only after office hours."

Pauli did not lecture well because he misjudged how much the crowd could take in, and his listeners did not often dare to interrupt him with questions. Once a brave student did interrupt, saying, "You told us that this conclusion is trivial, but I am unable to understand it." Pauli responded by leaving the room, something he often did when he wanted to think things over during a lecture. After a few minutes he came back and said, "It *is* trivial."

But on the whole, when students asked Pauli to explain something they didn't understand, he would do so with great patience and pleasure. We often said, "For Pauli, every question is stupid, so don't hesitate to ask him whatever you want."

In fact, Pauli loved people and showed great loyalty to his students and collaborators. All of his disciples developed a deep personal attachment to him, not only because of the many insights he shared but because of his fundamental endearingly human qualities. It is true that sometimes he was a little hard to take, but we all felt that he helped us see our weaknesses. Ehrenfest expressed our view well. When J. Robert Oppenheimer worked with him as a young post-doctoral scholar, Ehrenfest complained that Oppenheimer answered every question quickly but not always correctly, flying on to the next subject before Ehrenfest could correct him. Ehrenfest wrote to Pauli for help: "For the development of his great scientific talents Oppenheimer needs right now to be lovingly spanked into shape! He really deserves that treatment, in contrast to other young people, since he is an especially lovable chap!" Oppenheimer came to Zurich, but only for a short time, probably not long enough to get the Pauli treatment. Those of us who stayed longer were all spanked into shape by Pauli, and we loved it.

There was one person to whom Pauli reacted quite differently. When Arnold Sommerfeld, his former teacher, came to Zurich for a visit, it was all, "Yes, Herr Geheimrat, yes that is most interesting, although perhaps I would prefer a slightly different formulation. Might I formulate it this way?" For us, too often the victims of his aggressive tendencies, it was delicious to see this well-behaved, po-lite, and subservient Pauli.

His assistant had few regular duties. Pauli himself made up the homework problems for his class, and his assistant had only to grade the papers. My main duties were to be ready for discussions of his work and to assist him in any new developments. However, I once actually became his accomplice. His wife had asked him to change his eating habits in an attempt at reducing his famous girth. But Pauli loved sweets of all kinds, and many afternoons he would want to continue our discussions in a nearby *Konditorei* (café) that had won-derful pastries. I had to promise never to mention these clandestine outings to Mrs. Pauli.

Occasionally there were also more serious tasks. A lively correspondence was taking place between Heisenberg and Pauli about the problems of quantum electrodynamics. Some of these problems, such as the extremely large values for the mass and energy of the electron due to the very strong fields in the immediate neighborhood, were quite serious and were understood only much later. But many of them could be straightened out at the time. Whenever a letter came from Heisenberg, Pauli discussed it with me, and he frequently asked me to draft an answer. "You write it, I will correct it, and then we will send it to him," he would say. Once Pauli was more than usually dissatisfied with the contents of a Heisenberg letter. "Such silly statements. It is all stupid and wrong," he said. "You must tell him this in your letter." What could I do? I started the letter by explaining our disagreements as well as I could, and then I quoted Leporello in *Don Giovanni*: "My master wants to tell you, myself I would not dare to." Having absolved myself of any blame in this, I transmitted all of Pauli's curses.

At one point when he suffered from a severe depression, Pauli went to Carl Jung for psychoanalytic treatment. Through Jung he became very interested in various kinds of mysticism, including Jewish mysticism. This led Pauli to develop a friendship with Gershom Scholem, the world's greatest authority in that field and in the Cabala, the mystical system developed in the twelfth and thirteenth centuries. Pauli and Scholem saw each other frequently and exchanged their views in letters. Pauli rarely spoke about this side of his interests to his physicist friends. He did not discuss it much with me either, except that he urged me to visit Scholem when I went to Israel for the first time.

Meeting Scholem was a unique experience for me, because he introduced me to ideas that were totally foreign to those of our science. We had many remarkable and enlightening conversations. On one occasion Scholem asked me to tell him about unsolved problems in modern physics. I told him, among other things, about a certain relationship between fundamental constants whose origins we don't understand. For instance, one equation involves Planck's constant ($h$), the velocity of light ($c$), and the electronic charge ($e$).

The combination $hc/e^2$ happens to equal 137. My friend Val Telegdi used to point out that this number connects quantum theory ($h$), relativity ($c$), and electricity ($e$). Therefore, it holds a special significance for physicists. When I mentioned this number—137—to Scholem, his eyes popped out, and he asked me again if I had really said 137. He told me that in Hebrew each letter of the alphabet has a numerical equivalent and that the Cabala assigned a deep symbolic significance to the sums of such numbers in a given word. The number corresponding to the word *cabala* happens to be 137. Could there be a connection between Jewish mysticism and theoretical physics?

The day before my visit with Scholem I was in Safed with the Israeli mathematician Chaim Pekeris. Safed—now called Sfad—is the center of Jewish mysticism and the place where Rabbi Isaac ben Solomon Luria, an eminent interpreter of the Cabala, is buried. Pauli told me I should be sure to visit Luria's tomb, which is on the top of a hill in the cemetery. Unfortunately, we happened to have been in Safed on a Saturday, when it is forbidden by tradition to visit a cemetery. But since this was the only day I could visit the rabbi's grave, we decided to go anyway. Luckily, the cemetery was not locked. In order to show our respect, and perhaps because we thought it might generally improve the situation, Pekeris and I wore yarmulkes (skullcaps) when we ascended the hill to the tomb.

It was a splendid day with a clear blue sky and no wind. Then suddenly as we walked toward the grave we felt a gust of wind, and my yarmulke was blown away. We searched for twenty minutes and couldn't find it. We were both a little shaken, and Pekeris asked me whether we should proceed to the tomb. I said that we should—as scientists we shouldn't believe in such signs. We went to the tomb and on our way back searched again for the yarmulke, but in vain. It had miraculously vanished in the wind. When I told Scholem about this event the next day, he was delighted that a scientist had had a brush with the irrational.

Mystical and psychic experiences should be considered complementary to the scientific approach. Pauli's interest in these different avenues of human experience was in many respects a natural expansion of his involvement in modern physics. He was perhaps Bohr's closest disciple, being attracted to Bohr's generalization of complementarity, which was very close to his own thinking.

My time with Pauli was extremely rewarding. I did some of my best work under his guidance, and during the time I spent with him, I became a mature scientist. From Bohr, I received the inspiration and the philosophical attitude. With Pauli, I developed a thorough grounding in how to proceed in theoretical research.

Paul Scherrer was another physicist I met in Zurich who impressed me greatly. He was the director of the physics laboratory at the institute. His most important contribution to physics had come during his early days, when he developed a method of determining how the molecules assemble in solid material. At that time such determinations were being accomplished by scattering X rays from the material in the form of relatively large crystals. Scherrer, working with Peter Debye, found a way of doing this when the material was in powder form, which is much easier to deal with than crystals. The Debye-Scherrer technique is still widely used today.

Scherrer's interest drifted away from pure research to the development of one of the most active centers of experimental physics in Europe, and he became an inspired teacher—by far the best I've ever met. His two courses, one for beginners and one for more advanced students, were famous. Scherrer had a gift for devising inventive experiments that demonstrated the relevant facts and explained essential ideas memorably. His beginners' course was known as the "Cabaret Scherrer" because his explanations were often as witty as they were instructive.

The advanced course was even more remarkable because he could clearly present the most complicated of concepts. I was attracted to Scherrer's methods because of my own natural tendency to seek direct, easily understandable explanations. Like Paul Ehrenfest, who counterbalanced the mathematical formalistic approach that informed physics in Göttingen, Scherrer provided a refreshing contrast to the mathematical rigors that Pauli had imposed on the department in Zurich.

Often Scherrer and his assistants worked until late at night preparing his lecture demonstrations. Pauli also liked to work during the late evening. Once when he and I had discussed some difficult problems until nearly midnight, Pauli suggested that we should see what Scherrer was up to. We walked to the lecture room where Scherrer and his tired assistants were busy assembling a new experimental setup. Scherrer, full of enthusiasm, said to Pauli, "Listen to how I plan to explain the such-and-such effect to my students. You see, here is an electron with its spin up, and there is one with its spin down, and there they turn around each other and interact magnetically. . . . Isn't that simple?" "Simple it is," Pauli said, "but it is also wrong."

In spite of their differences about the value of easily understandable problems, Pauli and Scherrer were very fond of one another. When the Nazis took over Austria, Pauli was not yet a Swiss citizen, and according to Swiss law, his Austrian passport turned him into a German national. Staying in Switzerland would have been dangerous for him. (The Swiss granted Pauli citizenship only after his Nobel Prize in 1945.) When Pauli left Zurich for the United States in 1940, Scherrer accompanied him to the train station. Pauli returned the institute keys to Scherrer, who, with tears in his eyes, threw them onto the roof of the railroad car.

While I was in Los Alamos, there was some discussion about whether or not to ask Pauli, who was then at Princeton University, to join the atomic bomb project. He wrote a letter to Oppenheimer offering his services, if they were needed. We discussed this possibility but decided not to accept his offer. He would not have felt at ease as part of a large team. Although he had discovered nuclear spin, nuclear physics never interested him very much. We thought that it would be better for him to continue to work in fundamental physics so that the tradition would not get lost. I am sure that Pauli was very pleased by that decision. With his pure soul he would not have been happy working for destructive purposes.

When the war was over, Pauli was of great help to all of us. He gave a series of lectures at MIT about the latest ideas in particle physics, returning shortly afterward to Zurich. He wanted to keep in touch as closely as possible with American physics, and he asked me and others to write to him about the newest developments in theory and experiment.

In a letter I reported to him about the intention of the physicist Chien-Shiung Wu, at Columbia University, to test a suggestion by two Chinese-born theoretical physicists, Tsung-Dao Lee and Chen Ning Yang, that parity may not be conserved in radioactive processes. Conservation of parity is tantamount to saying that the mirror image of a process is also a possible process in nature. This is true for most natural phenomena. To put it another way, rotations to the left should be equally as probable in nature as rotations to the right.

Pauli wrote back that, in his opinion, this was a waste of time, that he would bet any amount of money parity is conserved in any process. Pauli's letter arrived just after I had heard the news that Wu and her collaborators, as well as other experimenters, had proved parity to be strongly violated. R. Garwin, L. Lederman and H. Weinreich and also J. Friedman and V. Telegdi had arrived at that astonishing result with several different experiments. The mirror image of a radioactive decay does not occur in nature. I am proud to say that my better self won out and I did not send a telegram to Pauli saying, "Bet for $1000 accepted" but reported to him by return mail the surprising result of these experiments. (Overseas phone calls were not yet used for physics at the time.) Pauli was completely flabbergasted. He wrote back, expressing his astonishment that "God is a weak left-hander." (Radioactivity is said to be caused by the so-called weak forces.) He added, "I am glad that I did not conclude our bet. I can afford to lose some of my reputation but none of my capital."

Pauli's style of theoretical thinking and research influenced physics all over the world. It emphasizes the essential roots and the symmetries of the laws of nature in their mathematical form and accomplishes this without much extraneous talk or hand-waving. His economical way of thinking and working remains for many of us an ideal to be emulated. Those who worked with him often ask, "What would Pauli say to this?" Sometimes we have to admit, "Pauli would never accept this."

However, Pauli, who died in 1958, set his example through more than the character of his work. He personified the striving for utmost clarity and purity in science and human relations. Pauli has often been called, quite correctly, the conscience of physics. We owe it in part to him that the community of physicists retains a certain amount of healthy simplicity, honesty, and directness in spite of all the politics, publicity, and ambitions—so foreign to Pauli—that now pervade the field. He was not only a great physicist; he was able to see deeper than others into both scientific and human problems. The dark riddles of the human psyche were not unknown to him. He remains an example to all of us of how to live a quiet and contemplative life of intellectual and moral integrity even in unruly times.

When Pauli was invited to America in 1935 for one semester during my time with him, he asked me to take over his course. This was the first time I had ever taught regularly. I admit that my confidence was strengthened by Pauli's deficiencies as a lecturer. Still, I spent a great deal of time preparing myself, and I was pleased and a bit amazed to see that the students seemed to like my teaching. To my even greater surprise and joy, I discovered that I loved to teach. Ever since then I have considered myself fundamentally a teacher rather than a researcher.

One day during Pauli's absence, I was called to the office of President Rohn of the Institute of Technology in Zurich. It was quite unusual for a person of my lowly rank to be called to the president's office, and I couldn't imagine why he wanted to see me. I entered his office, he invited me to sit down, and then without any pleasantries or preliminaries he asked me whether I had ever been a member of the Communist party or had any connection to the party. He looked at me sharply and said he wanted me to tell the absolute truth. I was taken aback and answered, "Mr. President, in all honesty I am not a member of the Communist party and have no connection to the party except that I do have a few friends who are members. But I want to assure you that I personally have never been in any way involved." That seemed to satisfy him, and he cordially issued me out of his office. Naturally, I was puzzled by this encounter.

A few weeks later I got a letter from the *Fremdenpolizei,* the police in charge of foreigners, summoning me to their offices. It occurred to me that this might have something to do with my talk with the president. I was shown into an office. On the desk I saw a large folder with my name and the ominous words *Kommunistische Umtriebe* (Communist Activities) written on it. After a few questions from the policeman in charge, I was shown the contents of the folder. To my amazement it contained a copy of every letter I had received since I had been in Switzerland. I was shocked that in this free country my mail had been opened, and when I later told this to my Swiss friends, many of them refused to believe it. The police must have gone to a lot of trouble copying each letter by hand or by typewriter, as there were no copying machines in those days.

The officer told me that they had found my correspondence contained a number of suspicious items. He said they would like to go over these things with me. First he wanted to know why I was friendly with so many European refugees, most of whom seemed to be leftists. I said, "I am a refugee myself, and I'm interested in these people as friends without necessarily sharing their political views." He seemed satisfied with this, but he went on, "There is the matter of your friend, Kurella. We have indications that you have close connections through him with the Communist party. You and Kurella lived in the same pension, and we have proof that you were very close to him, since his notebook contains repeated references to you, using your first name." I realized that they must have found Kurella's records of our card-game scores, but he didn't give me the chance to explain.

Instead, he continued, "We have a letter which you forwarded to Mr. Kurella after he was expelled from Switzerland." It was true that Kurella asked me to forward his mail to an address in Prague. "Do you know what this letter contained?" he asked me. I answered dryly, "I don't open letters that aren't addressed to me." Ignoring my remark, the policeman told me that the letters contained "highly political material" and added, "I can't believe that you didn't know what was in them." (Much later Kurella told me that these weren't important letters at all.) "There is more proof that you are close to the Communists," the interrogator continued. "Here is a letter from Professor Igor Tamm inviting you to come to Moscow and to stay in his apartment. Everyone knows that anyone who has a spare room in Moscow must be a big shot in the party." (Tamm was a well-known Nobel Prize–winning Russian physicist, but as far as I know he was not a member of the party.)

Finally, he pulled out a letter that I had written to my former sister-in-law, who was a member of the Anthroposophical Society. I tried to explain to this ignorant and simple-minded man that this was a religious group, not a political party, but nothing made any impression on him. I spent nearly the whole day in the police station going through this enormous pile of letters. When we were through, he said, "We'd like to expel you immediately, but unfortunately the Institute of Technology has vouched for you because they seem to need you. We will let you stay here until the term is over, but then you will have to leave Switzerland unconditionally. Any appeal is useless. You and your wife will have to leave the country, and you will never be allowed to return." I realized then that the president of the institute had come to my defense after our conversation in his office. Many years later when I was asked to be director general of CERN (the Conseil Européen pour la Recherche Nucléaire) in Geneva, I showed them the letter excluding me from Switzerland for all time. Of course, they laughed it off.

In 1935, however, the *Fremdenpolizei* was no joking matter. Some of the officers were Nazi sympathizers. I remember a young student who worked with me, a German refugee named Eichenwald, who was called to the police. They asked him the same kind of questions they had asked me: Who do you see, what groups do you frequently meet with, where do you go in the evenings? Eichenwald said, "I see mainly Pauli, Weisskopf, and other professors." The policeman replied, "That is typically Jewish. When I was in college I never spoke to any professors."

Switzerland is a small country, and there was a tremendous stream of refugees trying to get in. Many had walked across the frontier. But there were no guarantees that having safely reached Switzerland they were out of danger. If they were caught within fifty kilometers of the border, they were frequently sent back to the Nazis. If they were found more than fifty kilometers inside Switzerland, they were assigned to detention camps and required to do manual labor. The Swiss had very little room in their small country and apparently not enough sympathy for the refugees. After the Nazi defeat at Stalingrad in December 1942, the Swiss radically improved their treatment of refugees. (The Swiss film *The Boat Is Full* describes the dark side of this story very well and gives credit to those decent Swiss who harbored the refugees illegally in their homes.) One famous case involved a liberal journalist who was somehow able to escape to Basel, on the German frontier, in 1942. He came without permission but was able to live in Basel until one day a gang of Nazis recognized him, dragged him into a car, and brought him back to Germany. The Swiss were very angry about this and accused the Germans of illegal acts. They eventually forced the Germans to return him to Basel, but then the Swiss expelled him immediately to France.

# SIX

❖

# The Growing Threat of the Nazis

THE INTERVIEW with the police had no effect on my life, and I continued teaching. But then Pauli returned from America and my course was over. It was time to move on. With some regret, Ellen and I prepared to leave Switzerland. We had had a wonderful time there, in spite of personal experiences that proved world politics could reach even into this seemingly isolated country. We had to make plans for our next position. But where could I find a job in Nazi-ridden Europe? Fortunately, Niels Bohr was actively helping people in my position by collecting money from many sources to help refugee scientists. He had been able to arrange a scholarship for me from the Carlsberg Beer Company. So we returned to Copenhagen and the Bohr Institute, where we found the best of Europe's refugee scientists, including old friends like James Franck, George von Hevesy, Robert Frisch, Hans Halban, George Placzek, and others. It was wonderful to be back in Copenhagen and to join once more the illustrious collection of physicists inspired by the great master Bohr.

Influenced and stimulated by this remarkable group, I wrote two of my best papers during that time. One was on the properties of the vacuum in the new quantum theory of the electromagnetic field. The vacuum plays a special role in quantum electrodynamics—it is not considered a completely empty entity, as first conceived. In chapter 5, I mentioned pair creation: when enough concentrated energy is available and a pair consisting of a particle and an antiparticle is produced out of empty space. But even if such energy concentrations

95

are not available, the vacuum behaves in a special way, according to quantum field theory. For example, an electric charge, such as an electron or a proton, induces in the vacuum a cloud of opposite charge surrounding the original charge. This so-called vacuum polarization was a topic that caught my interest from the beginning. The induced cloud is rather weak in charge, but it has indeed been observed in nature in the neighborhood of the original particle, through delicate experiments. The effective charge of the proton becomes larger when one measures it inside that cloud. The discrepancy between the actual charge at the center of the cloud and the one observed at greater distances is one of the curious features of quantum electrodynamics.

Not only the charge but also the mass of the particle becomes large when the effects of fields in its immediate proximity are taken into account. In my paper I remarked on these effects and proposed to set aside the effects at extremely small distances, simply by postulating that the total charge (intrinsic plus vacuum cloud) and the total mass (intrinsic and that caused by the field effects) are what we find in nature. I could show in the paper that all other effects of the interplay between charged particles and the electromagnetic field, except the mass and the charge, should not be influenced by the fields in the immediate neighborhood of the particle. Thus, it would in principle have been possible to predict some small but interesting consequences of that interplay between particles and fields, such as a slight shift of the energies of the states of the hydrogen atom—the so-called Lamb shift—and a small change of the magnetic properties of the electron. These effects had not yet been observed at that time or were only vaguely suspected. All this was developed in great detail after the war and led to a successful reformulation of field theory.

There were no fundamental reasons why I could not have calculated these effects, which were discovered ten years later. I just was not clever enough and perhaps not interested deeply enough to indulge in the long and tedious calculations necessary. Also, in 1936 my interests were too broad, and I was not eager to delve into the detailed consequences of my own work. I was deriving so much pleasure from understanding other developing aspects of the new physics such as the growing knowledge of the structure of the atomic nucleus—that I could not continue to concentrate on just one area of the field. If I had, I might have succeeded in recognizing these effects earlier than anyone else. But that would have deprived me of the pleasure of exploring the vast horizons of the new physics I saw emerging.

The other paper I wrote at that time was influenced by new insights into the structure of atomic nuclei, spearheaded by the innovative ideas of Bohr and his collaborators. It dealt with the effects of energetic particles colliding with nuclei. I tried to describe these effects as a rise in nuclear temperature followed by an "evaporation" of neutrons. This idea had also been promulgated by Yaakov Frenkel and by Lev Landau in a different way. It was interesting to apply thermodynamic concepts to objects as small as atomic nuclei. My paper contributed to the understanding of what happens in nuclear reactions. A nuclear reaction is a process in which an energetic particle impinges upon an atomic nucleus and delivers part of its energy to the nucleus, which in turn emits other particles and is transformed into another nuclear species. The first nuclear reaction was performed by Ernest Rutherford in 1919. He bombarded nitrogen atoms with energetic alpha particles, which are emitted by radioactive substances. When the nucleus of the nitrogen atom was hit by an alpha particle, it emitted a proton and was transformed into an oxygen nucleus, which formed an oxygen atom. This was the first time one element (nitrogen) was transformed into another (oxygen). The alchemists' dream had been realized. Later on many other nuclear reactions were found.

With the publication of this paper I became a member of the community of nuclear physicists. I stayed with this field until 1961. However, I did not leave quantum field theory altogether. I published a number of papers designed to clear up some of the problems involved with the interaction between electrons and radiation. The subject's intrinsic interest was not my only reason for writing about nuclear reactions. I knew that nuclear physics was expanding rapidly in America, and I thought my work in nuclear physics would help me find a job there. I was eager to publish the paper in the American journal *Physical Review,* and I sent it off before Bohr had a chance to see it. At the same time I sent a copy to Bohr, who happened to be in America at that time.

A short while later I received a letter from Bohr's collaborator, Fritz Kalckar, who had accompanied Bohr on his trip, reproaching me because I did not sufficiently emphasize Bohr's ideas as the inspiration and basis for my paper. The letter was quite damning. Kalckar pointed out how much Bohr had done for me and added that he deserved better treatment. This was my first and only conflict with Niels Bohr. I was, of course, deeply upset and flustered because I had thought that I sufficiently cited Bohr's contributions. I recalled the paper for reformulation. When Bohr returned from America, he suggested only one or two changes, none of them substantive. They were so minor that I couldn't understand why he had been so upset about the original formulation. But I was relieved that my personal relationship with Bohr, which was so important to me, had not been damaged.

Every year in September, Bohr convened a conference to which he invited his previous collaborators and other prominent physicists who worked at the frontiers of science. I was fortunate to have attended these annual events ever since my first stay in Copenhagen. They were an impressive testament to the influence Bohr exerted on the worldwide physics community. All the newest ideas and experimental findings were presented and discussed in detail. Bohr led the discussions and always asked exactly those questions that illuminated the significance of the subject under discussion. Most participants went home from these gatherings with new ideas and projects. These meetings were attended by not more than about fifty participants. This fact made them much more stimulating than similar conferences today with several hundred people.

In addition to the serious purposes of the conferences, it had been a tradition since 1930 to hold an evening devoted to what we called "comic physics." After I began to participate, I joined a small group consisting of Hendrik Casimir, Carl Friedrich von Weizsacker, and George Gamow, who assumed the responsibility of preparing an evening of skits and parodies aimed mostly at the older generation. One evening's program (before I took part) was a parody of Goethe's

*Faust* devised by Delbrück and Weizsacker, in which Niels Bohr was God the Father; Ehrenfest, Faust; and Pauli, of course, Mephistopheles. Ellen played Faust's sweetheart Margarete (who represented the neutron). Her song, a takeoff on the famous line "Meine Ruh ist hin, Mein Herz ist schwer (My repose is gone, my heart is heavy)" went something like this: "Meine Ladung ist hin, statistics ist schwer (My charge is gone, statistics are hard)." This parody is still considered a classic by those who were in the audience as well as those who have only read about it.

We took our responsibility for the comic session very seriously, working during lunch and on evenings when there were no scientific discussions. We rewrote the lyrics to popular songs, substituting jokes about science that only physicists would appreciate. One year, because Bohr had been on a trip around the world, we used Jules Verne's *Around the World In Eighty Days* as our text. Bohr was Phileas Foggy, a reference to Bohr's somewhat foggy way of speaking. He was supposed to have met strange animals on his voyage. A rhinoceros resembled Pauli, a gazelle talked like Heisenberg, and a little monkey was perhaps meant to be Marcus Fierz, a Swiss physicist. We wanted to find an animal that reminded us of Placzek, but we couldn't—so we invented "the Placzek." It was a pleasure to see Bohr, Pauli, Heisenberg, Meitner, and Ehrenfest in the first row during these performances, obviously enjoying the jokes directed at them.

I was constantly amazed, during that time, by Bohr's apparently limitless energy. He was deeply involved in science, writing some of his most important papers about nuclear physics and electrodynamics. In addition, he worked hard to get money from Danish, American, and British sources for "his refugees" who could no longer work in their own countries. Every year he traveled to England and America to find us jobs.

In the fall of 1936 I was offered a full professorship in Kiev, which promised a good salary and opportunities for yearly trips abroad. At the same time, I was also being considered for a job as a theoretical physicist at the Moscow laboratory of Professor Peter Kapitza, one of the world's foremost experimental physicists, whom I knew from Cambridge, England. In 1934, on one of Kapitza's visits home from

England, Stalin had revoked his exit permit, forcing him to remain in the Soviet Union. As a concession, however, Stalin subsequently agreed to ship all of Kapitza's laboratory equipment to Moscow.

In view of his interesting and inspiring personality, a job in Kapitza's lab would have been a most attractive option for me. Apparently, before offering me the position, Kapitza had asked Dirac for advice about whom to appoint. Some letters from the correspondence between Kapitza and Dirac were published in the January–February 1990 issue of *Science in the USSR,* the magazine of the USSR Academy of Science, and they included a letter from Dirac recommending me to Kapitza. This letter was new to me. Dirac's well-known reticence regarding making any statements about other people makes the letter a rare exception to his ordinary behavior, and so I want to quote it.

October 30, 1936

Dear Peter,

Thanks for your letter.

Weisskopf has been working mostly on problems of a general theoretical nature and has shown outstanding ability. Some time ago he collaborated with Pauli and wrote a paper which has become quite famous as the Pauli-Weisskopf paper. This paper shows that one can construct a relativistic quantum theory for particles without spin (which was previously believed to be impossible) when one takes into account the possibility of the creation of pairs. This work does not have a practical application, since fundamental particles of zero spin are not known, but it plays a most important part in all theoretical discussions of the ultimate particles of nature. More recently W has been doing valuable work on the forces in nuclei and when I was in Copenhagen I saw his latest work on the polarisation effects which occur in the theory of the positron. This is a very good piece of work—it provides a more physical way of dealing with the question than Heisenberg and I have done.

In my opinion Weisskopf is the most suitable man for you. I had forgotten him when we were discussing the possibilities together. He is a more powerful theoretician than Guerney and is about equal in this respect with Heitler or London. Weisskopf is the most agreeable man personally (and has a very charming Danish wife). Also Weisskopf has already been in the USSR (he was lecturing in Kharkov, I think) and likes it there and has learnt some of the language. . . . W has been urged to take a position in Kiev. . . . [He] is going to the USSR . . . and will call and see you on his way through Moscow. I gave him the address of your institute.

I accepted an invitation from Kiev to go to Russia, to talk things over both with the Kiev people and with Kapitza. I did it because it gave me an opportunity to observe firsthand the changes that had taken place under the steadily increasing terror of Stalin, and I was eager to take Ellen on her first trip to the Soviet Union. As it happened, George Placzek was invited to the USSR at the same time, so we planned to travel together.

Placzek made a list of things that were hard to get in the USSR to bring to our Soviet friends. Among many other items he decided to include a large quantity of condoms. Since he had a very generous nature, he bought about a thousand of them. We decided to assemble all the goods in our apartment. On the appointed day, Placzek arrived with his arms loaded with packages. He began to unpack the boxes of chocolate, slide rules, tissues, and soaps he had bought. Then he suddenly stopped, hit his forehead with the palm of his hand, and said, "The Copenhagen taxi trade is going to die out!" We couldn't imagine what he meant. "I forgot the box with the condoms in the taxi," he finally told us, and he left to buy another huge batch.

We left for the USSR without any illusions, but what we found was even more dispiriting than what we had expected. We were instantly and sharply aware of an atmosphere of fear and terror. Some of the friends we called acted as if they had never met us. (Among these was my old friend Sasha Rumer, who was put into a camp not long after our visit.) We found that most people were afraid of contact with foreigners, except on official or professional business such as the talks we delivered at the university or in research institutes. If I had had any doubt about it, I was quickly convinced that a job in Russia, even under the most generous of terms, would have been most problematic. During that time, the winter of 1936, the fear was most apparent in Moscow; it had not yet penetrated into the provinces. In Leningrad, Kharkov, Kiev, and Odessa, we received a much more friendly reception.

The story of the Kapitza offer had a strange ending. There is another letter in the Dirac–Kapitza correspondence that Kapitza wrote to Dirac on 17 February 1937, in which he said, among other things:

> About the theoretician, I met Weisskopf, and he said that till autumn he is staying with Bohr and then, if he does not find any other prospects, he is quite willing to come. I hate people saying that if they find nothing better they will come here.
>
> Meanwhile Landau said that he [would like] to come here to work with us and I consented to take him. He is as good as any other theoretician who wanted to come, but being Russian I shall have less trouble with him than with a foreigner.

Actually, Landau was much better than any other candidate, including myself. But Kapitza's jab at me was perhaps unjustified. I was too well acquainted at that time with the situation under Stalin; I would have accepted a position in the USSR only if I had had no other possibility. Kapitza himself had experienced repression under Stalin; he should have understood my reluctance to accept a job in the Soviet Union.

After our return from the Soviet Union in the early spring of 1937, Bohr came back from one of his trips to America. Now it was my turn; Bohr convinced the University of Rochester to offer me an instructorship. I was grateful, but the salary was rather low compared to what I had expected to get in Europe. Still, I quickly accepted the offer. I have never felt any regrets about that decision.

Indeed, the experiences of some of my friends in the USSR confirmed the rightness of my decision. Shortly after our visit Alex Weissberg was arrested in Kharkov. One of the reasons for his arrest may have been a dangerous joke played by Placzek during that trip. In Kharkov Placzek was offered a job at the Physico-Technical Institute. During an evening party he said he would accept that offer if five conditions were fulfilled: first, a salary of a certain amount;

then part of the salary to be paid in dollars; then every year a visit to the West; then a certain number of collaborators; and lastly, "the *khasain* must go." The word *khasain* means something like "leader" and referred to Stalin. Unfortunately, a faithful Communist who was at that party reported this to the authorities. Placzek had to leave Russia immediately.

Soon after this, Weissberg, who had invited Placzek to the Soviet Union, was arrested. He was held in detention for many years without being told of the crimes he was supposed to have committed. Many times he was interrogated and often tortured to force his "confessions" or to provide his captors with names of supposed accomplices. He was able to resist the pressures and later had the bitter satisfaction of meeting some of his interrogators in the detention prison—even they had been arrested as the terror spread. After several years in prison Weissberg was released, but he was sent back to Nazi Germany, which was tantamount to a death sentence for a Jew. He wrote of his experience in the Soviet prison camp in a book called *Hexen Sabbat,* which appeared in America in 1951 as *The Accused.* He told me the rest of his story when I met him much later in Europe.

Weissberg was an indestructible man. The Nazis shipped him off to one of their extermination camps in Poland, but on the way, the truck in which he was being transported broke down. He jumped off and disappeared into the woods, dodging the guards' bullets. After wandering among the trees for some time, he stumbled into a camp of Polish partisans. A certain Countess Cybulski, whose husband disappeared in the West, was a member. Alex married the countess and soon became the leader of the group. His exploits in the underground had an almost miraculous quality, and his life took on the characteristics of a movie or a popular novel. He traveled several times to Switzerland in order to get weapons for the underground. He was in the Warsaw ghetto during the uprising. The Nazis never succeeded in recapturing him.

When the war ended, Weissberg worked in the commerce department of the new Polish government, organizing the export of timber. With great misgivings he witnessed the triumph of communism in Poland. After a lucrative business trip to Stockholm he decided not to return to Warsaw. In his next career he became a successful independent timber dealer. He and the countess moved to Paris—where I visited him and heard the incredible story of his life—and he continued to be enormously successful in business not only in timber but also in railroads. He built a railroad somewhere in South America. Unfortunately, he died before he could write a book about his further adventures.

Sasha Rumer's life was more tragic. After his arrest he was sent to Siberia. He survived mainly because the guards needed a teacher for their children, and he was the only educated person in the camp. In the meantime, his colleagues in Moscow, mainly Landau and Kapitza, tried to get him freed. It took them nearly ten years, but Rumer was freed before Stalin's death. As he told me later, the worst came after he was allowed to leave the prison. He was not permitted to return to Moscow and had to stay in Siberia. He and his wife tried to find an apartment, but in the fearful atmosphere then pervading Russia no one dared rent to a released political prisoner. So they lived in cellars and often had to move precipitously from one place to another.

After Stalin died, Rumer's "criminal record" was expunged. He was completely rehabilitated and told he would be allowed to go back to Moscow. Instead, he chose to live in Novosibirsk, where he could teach and do research at that center of science. I met him on one of his frequent trips to Moscow and later during a visit in Novosibirsk. He was a leading member of the Siberian Academy of Science and led an agreeable and productive life until his death in 1985.

Ellen and I felt tremendously relieved to return to the West shortly before Christmas of 1936. After the tense weeks in Stalin's Russia it was a great joy to arrive in the midst of Stockholm's welcoming, beautiful pre-Christmas atmosphere.

In Copenhagen, we prepared to leave Europe. The ominous signs surrounded us; the rise of Nazi power, the persecution of the Jews, and the Spanish Civil War made it clear that a European war was not far off. It was time to go. We wanted to be as far away from the oppressive and murderous Hitler regime as possible. Luckily, none of my close relatives or acquaintances were among the victims of the Nazi persecution. My family and most of our friends were able to leave Austria when Hitler occupied it. But some of my father's relatives from the Jewish community of Sucice were caught. My father's niece, Mitzi Klein, married the poet and writer Fritz Baum, and they moved to Prague. Unfortunately, Fritz Baum was lame, which prevented them from leaving Czechoslovakia when the Nazis came. After the war we found their names and those of other relatives of my father's among those of the Nazis' victims engraved on a wall in a synagogue in Prague.

Somehow I lived a charmed life during the Nazi era. I was never personally attacked or mistreated. Nevertheless, that period of German depravity left an indelible scar on everyone who lived in Europe at that time. It did not destroy my intellectual and emotional bond to the best of German culture, which was and is deeply ingrained in my thinking and feeling, but it demonstrated that unfathomable evil can take hold of even a civilized society.

We spent the summer in Austria with my family and bid a sentimental farewell to Vienna and the mountains in Altaussee. We did not know if we would ever see those familiar, beloved views again. At the Salzburg music festival we heard *Der Rosenkavalier*. The next day, as we were walking by the opera house, we saw the scenery being loaded onto a truck to be shipped away. To us at that moment it seemed a symbol of the end of Austrian culture. We were afraid that the Salzburg festival and Austria's glorious music were things of the past and that the coming Nazi onslaught would destroy the culture that had been such an important part of our lives. With heavy hearts we returned to Copenhagen. In September 1937 we left for America on the Swedish liner *Gripsholm*. Along with our few possessions we took a great deal of hope and concern for the future.

# SEVEN

⊠

# Becoming an American

WHEN WE ARRIVED in America, we were met by our old friend Max Delbrück. As always, Max was the perfect welcoming committee for the new arrival. With his customary enthusiasm he greeted us by saying, "You're going to love America, especially New York. It is absolutely wonderful." Then he proceeded to explain the Manhattan street system: the avenues ran from north to south and the streets from east to west. He told us about the two rivers and proudly explained the remarkable concepts of "uptown" and "downtown." Eagerly he described his favorite restaurant, Schrafft's, where you could get sandwiches and a fabulous delicacy, apple pie, cheaply. But we hardly had time to sample New York because in a few days we took the train to Rochester, our new home.

The head of the University of Rochester physics department was Lee Dubridge, a well-known scientist who ran a small but very good department. We were overwhelmed by our reception in our new home. In Copenhagen and Zurich I had been considered a foreigner and therefore somewhat outside of the social circles. In Rochester it was our foreignness that helped draw us into the social life—everyone found us interesting. In European countries an accent was considered a disadvantage, but in America even our accents were an asset, a glamorous advantage.

I learned that I had been given the job because another physicist, Fred Seitz, had left for a new position in Philadelphia. Leonard Schiff, who had just finished his Ph.D., had also applied for the position. When Schiff heard I had been hired, he wrote a letter to Dubridge in which he regretted that he could not get the job in Dubridge's department, but expressed his appreciation that the job would be given to me, for two reasons: First, because he knew how much a refugee from Hitler would need a job, and second, because of the enrichment of the American physics community by my presence. I was very touched by his letter. Leonard Schiff became an outstanding physicist, but he died young and a lectureship was established in his name at Stanford. I am proud to have been asked to give one of the memorial lectures.

The University of Rochester paid me a salary of only $200 a month, which was just about enough for us to live on. After several years I was promoted to assistant professor with a raise to $225. There is no doubt that some refugees were exploited when they came to America. Like me, many immigrant scientists and other professionals were paid less than the usual salaries for people of their experience. But we had no choice, and the warmth of our reception made up for the low salary.

During my Rochester days, Robert Serber, a physics professor at the University of Illinois at Urbana-Champaign, left his job to accept a position at Berkeley. Because I knew Wheeler Loomis, the head of the Illinois department, I was certain that he would offer me Serber's slot. When I heard that he had chosen someone else, I was stunned, but also puzzled. Later, during the war, I saw Wheeler at a cocktail party and said, "I've been wondering for years why you didn't hire me when Serber left." Wheeler didn't hesitate: "That job was only an associate professorship without tenure, and I was convinced that you had a tenured full professorship at Rochester." He couldn't imagine that Rochester would have offered me anything less.

Nevertheless, our stay in Rochester was extremely pleasant in all respects. My work was interesting, and the people were very friendly. But to the question, "When will you have children?" we always answered, "First we must have parents," meaning that our top priority was to get my mother and the rest of the family out of Europe and away from the Nazi threat.

Within half a year after our arrival in America, Hitler had taken over Austria. My mother, brother, and sister were still in Vienna.

(I have often thought that it was lucky that my father died before the Nazis took over. He would never have been able to face emigration from Vienna, where he was so successful and well known.) I had warned my family in the months before we left that they should try to get some of their money out of the country because we would all have to leave soon. They didn't believe me, and besides, they always said, the interest rate was much better in Austria.

I managed to overcome their reluctance to some extent. Whenever I went to visit them, I took the maximum amount of cash allowed back to Copenhagen and deposited it into a Swiss bank. One time my mother must have counted wrong. When I got to Copenhagen I found that she had given me ten times the legal amount. Luckily, I hadn't counted the money before leaving Vienna, so I didn't realize what I was doing. I simply declared the usual amount when I got to customs. I would never have been able to smuggle the money across the border if I had been aware of the amount I was carrying. Fortunately, the officer didn't ask to see the money, or I would have been arrested.

I was only able to get out a total of $10,000 (including the illegal sum), and that was certainly not enough for my whole family to live on for any length of time. In order for them to enter the country without a job offer, they needed an affidavit signed by a U.S. resident assuming financial responsibility for them. My low salary and lack of savings made it impossible for me to sign such an affidavit for more than one person. Although we still owned considerable real estate and capital in Vienna, the Nazis not only prohibited the export of money but also charged would-be emigrants a sum equivalent to their entire holdings as a fee for their passports. My mother had to sell all her real estate for a fraction of its worth.

I was also worried for George and Annie Winter and their little son. George, the boyhood friend with whom I studied the stars, had been unable to find a job in the West because of the economic depression in the early thirties, and had worked in Sverdlovsk as an engineer. In 1938, when conditions became intolerable there, the Winters left Russia, arriving in Vienna only a few weeks before the Austrian-German *Anschluss*. After that, because they were Jewish, Austria was even more dangerous for them than Russia.

It began to appear impossible for me to bring my family to America, until I contacted several Jewish organizations in Rochester that were extremely helpful. Soon, through the help of a generous and wealthy Jewish philanthropist, Jesse Horwitz, I had affidavits for my entire family as well as for George and Annie, who became known as Anne in America.

When they arrived in America, my mother came to live with us in our home in Rochester. My brother could not resume his career as a lawyer but got a teaching position in economics at the University of Omaha. My sister taught psychology at Briarcliff College in upstate New York. My mother, then fifty-eight, found the transition to a new country, a new language, and many unfamiliar customs very difficult. Because we were in our early thirties, we had far less trouble adjusting to a new life, but she found it nearly impossible to manage. Ellen and I did what we could to help her settle into life in America but she never really felt at home in the States. She died in 1958.

After my family and our closest friends had been brought over, it was time to start a family of our own. We had always wanted to have children so that our ideals could survive us. Ellen and I felt confident and ready to become good parents. Our son was born on 13 April 1940. We called him Thomas because of our admiration for Thomas Mann, the German novelist who had been expelled by the Nazis for his political opinions.

By now the Nazis had occupied Ellen's homeland, Denmark, and we were filled with sorrow and fear for our friends. We could not have imagined in those dark days that only five years later little Tommy would be in Los Alamos celebrating the end of the war. In 1942 our daughter, Karen, was born. At that time her name was rare in the United States, but we chose it because of its Scandinavian origin. It was also the name of Ellen's sister.

While my mother lived with us we asked her to help take care of the children. My mother had many strong and sometimes old-fashioned ideas about how to bring up children and that led to some conflicts between us. The usual conflicts between the generations can be especially hard to take when children welcome older parents into their home. But we had no choice in the matter. My mother had to

109

leave Austria, and there was no other place for her to live. We tried to make the best of it.

We had many good friends in Rochester. There was a small group of liberal people who wanted us to join their club, which met every month in the home of a member. We jokingly called ourselves the Red Club, although none of us were Communists. These monthly dinners made us feel at home in Rochester among our first group of American friends.

I was the only physicist in the club until a young German graduate student, Gerhard Dessauer, joined us. There were several historians, two staff members of the museum of fine arts in Rochester, a political scientist, and a professor of religion. This diverse group helped bring Ellen and me into contact with many sides of American culture. Our discussions included U.S. politics—we all enthusiastically supported Franklin Delano Roosevelt—and America's potential role in the war. Ellen and I gave eyewitness accounts of the Nazi and Stalinist terrors, which aroused great interest. Our reports about Soviet totalitarianism made a particularly strong impression, since many of these liberal thinkers felt that the stories in the newspapers were biased against the USSR.

Making friends in America was much easier than it had been under similar circumstances in Europe. People seemed more ready to give advice and to help newcomers. Most of these friendships may not have been as deep as our European ones, but some of them certainly were. Among our closest friends were Sidney and Peg Barnes. Sidney was an experimental physicist who taught at the university. He came from a midwestern farming family and had a completely different background from us. Some would have suggested that he was not an intellectual, but they would have been wrong. The Barneses did have a different outlook from the people who comprised the central European intellectual milieu of our past, but we had a wonderful time together and learned much from one another. Ellen and I owe a good part of our understanding of the American way of life and thinking to Peg and Sidney.

Another close friendship developed with Elva and John Richards. He was a classical philologist at the university. Because he was British and she was Swedish, they shared some of our European prejudices.

On the whole, however, we had little contact with other European refugees. We wanted to establish ourselves in the American milieu. Our conscious efforts in those first years to establish personal ties with American life led to a reputation we cherished very much. Some Americans told us we were more American than most of the other refugees. Paradoxically, when we later returned to Europe, our friends there told us that we retained more of our European character than most refugees.

Under these favorable circumstances we completely avoided the culture shock that many refugees experienced when they came to America. We were glad to have left the turmoil in Europe, and with open minds we readily adjusted to a different way of life. After all, academic communities all over the world were then, as now, similar to a certain extent. It was easy to find some new friends whose cultural attitudes were similar to ours, and we did not need to change the basic precepts that guided our lives.

The Rochester department was small, consisting of a teaching staff of about a dozen people and fifteen or sixteen graduate students. The relationships between professors and students were very close, and there was none of the formality that characterized European academic life. A lot of the students were close to our age, and we spent many happy, relaxed evenings with them, often at our house. It was quite intimate. We discussed physics, but we also talked about their personal problems, as friends would.

Some of these students eventually became very important in their field. Herbert York discovered the uncharged pi-meson when he was at the University of California in Berkeley. He became undersecretary of the air force under President Jimmy Carter. An advocate of arms control, he is now at the University of California in San Diego. Robert Dicke, now at Princeton, has become one of the world's leading physicists and cosmologists.

Because we were so interested in American culture, we spent what little money we had on trips to New York, where we went to plays, concerts, and museums. Through the initiative of J. Robert Oppenheimer, whom I had met some time earlier in Europe, we were introduced to a very interesting group of people in New York. They were close friends also of his brother Frank and usually assembled in the home of Charles and Mary Ritter. Charles was a designer of sound equipment, mainly for the blind. The group members were interested in everything that was happening in the arts, science, and politics. On the whole, they were much more radical and left-leaning than the members of our not-so-Red Club. Ellen and I felt very much at home in this group, which reminded me of my Vienna days in the socialist youth movement. To my delight they even sang the familiar songs of my youth in English translation. When we went to the political cabaret called Pins and Needles sponsored by the International Ladies Garment Workers Union, I was reminded of my cabaret days and our more amateurish political parodies.

Of course, we also attended physics conferences and the meetings of the American Physical Society, which took place on university campuses with just a few hundred attendees and, unlike today's meetings, were intimate affairs at which one knew most of the participants. After the meetings, at which the important developments in science were discussed, we would all go out for an evening of fun and conversation. Especially congenial were the gatherings of theoretical physicists at Edward Teller's house in Washington, D.C. At that time Teller was extremely popular and sought after for the warm atmosphere in his home and his readiness to discuss any physics topic one could think of. Those were the good old days of seemingly complete political agreement, when everyone was united in their concern about the one overriding threat to the world, the Nazi regime in Germany. We made all those trips by the cheapest possible means, by car or bus. There was no department support in those days for travel to scientific meetings.

In the 1930s nuclear physics was the most fashionable of all the branches of physics. Because of my work with Bohr in Copenhagen, I was considered an expert in nuclear structure. Sidney Barnes was just starting to build a cyclotron (a special accelerator using both electric and magnetic fields) in Rochester; the device produced beams of protons at an energy of up to six million electron volts, which was very high for that time. I was soon deeply involved in planning the program. With the help of a graduate student, Douglas H. Ewing, I calculated the expected rate of nuclear reactions induced by the proton of our cyclotron. (Ewing was the only one of my many collaborators who succeeded in putting my name ahead of his in our paper. He surreptitiously changed the order of names in the galley proofs and made me break my long-standing alphabetical vow.)

At that time the radiation of light by nuclei was a subject that interested me. When nuclei get excited by bombardment with energetic particles, they tend to emit light rays. I devised a formula that gave an indication how strong this radiation would be, depending on the character (multipolarity) of the radiation. The intensities resulting from this formula are often referred to as "Weisskopf units."

I tried to keep my physics interests broad and also worked on what today would be called "particle physics." I was helped in this by Robert Marshak, who got his Ph.D. with Hans Bethe at Cornell and then came to Rochester as a postgraduate student. We discussed certain strange properties of mesons, short-lived particles produced by cosmic rays when they penetrate into matter. We published a few papers together that pointed to the problems rather than explained the properties. After the war Marshak found the solution to the problems.

Marshak was the son of a poor Jewish immigrant who came to America from Russia during the pogroms of 1905 and earned a meager living as a tailor. His is a success story typical of the son of Jewish immigrants. Academia was not free of anti-Semitism before the war: When the physics staff examined applications for graduate students, Jewish students were accepted only when they had exceptional recommendations, and even then not always. Therefore, those who got in were on the whole better than the other students. I became aware of this when we expressed our long-standing wish to move to the Boston area. We were told that it was very difficult for a Jew to get a position at Harvard or MIT. (All this has almost completely disappeared in the postwar period, and anti-Semitism has practically vanished from academia.)

I had recommended that Bob Marshak be appointed an assistant professor at Rochester in 1943, but it was not easy to convince the authorities to give him the job. When he was finally hired, he wanted to introduce me to his father in New York. We entered the modest apartment, and I met an old man with a long beard who spoke mostly Yiddish. When he was told who I was, he tearfully embraced me and said, "My whole life with all its deprivations was worthwhile; my son is now a professor and you helped him do it." Marshak became head of the department at Rochester after the war and later was president of the City College of New York. He and his collaborators made many important contributions to our knowledge of the physics of elementary particles.

At Rochester, I taught classes and began to build a reputation as a lecturer. It gave me great pleasure to organize my courses so that the students understood what I was trying to say. I found that if I couldn't explain a subject clearly, it was a sign that I hadn't really understood it myself. Consequently, I spent a great deal of time thinking about my courses, and the students seemed to appreciate my efforts.

In the summers of 1938 and 1939 I was invited to teach at Stanford University. We drove across the continent, a great experience for us. There were no superhighways in those days, so we saw a great deal of the country intimately. Once in the badlands of North Dakota we got stuck in the mud in a wild area where there were no paved roads at all. Placzek came with us on one trip. In Copenhagen in 1934 we had made a bet about how long Hitler would last. Placzek said not more than five years, and I said he would be around for much longer. At first we wanted to bet fifty dollars, but Placzek thought the value of the dollar was subject to change. Instead, he proposed, "Let's bet fifty Wiener schnitzel." At the time a Wiener schnitzel cost about a dollar. By the time we were on our way to Stanford, I had unfortunately won the bet. When we stopped at a restaurant that served Wiener schnitzel, Placzek threatened to pay off what he owed me by ordering fifty, but I managed to persuade him to pay me the money instead.

While we were out west, we visited Oppenheimer on his ranch in New Mexico. It was a wonderful place in the Pecos Valley near Los Alamos, which at that time was known to us only as the site of an obscure boarding school for boys. There was always a nice group of people at the ranch, and everyone pitched in with the cooking and the household chores. The ranch had few conveniences and no electricity, and we slept out on the terrace. Placzek, a confirmed city dweller, said he couldn't sleep because chipmunks danced on his stomach all night.

In the summer of 1938 we had conversations with Oppenheimer that lasted all day and into the night. Oppie, as we called him, still believed to a great extent in communism. We tried to convince him of the reality of Soviet life by describing the lack of freedom and the persecutions under Stalin that we had seen there. I believe that these conversations had a lot to do with his eventual rejection of communism. He trusted us as stalwart liberals who were not ideologically anti-Soviet but who had had an opportunity to see life in the Soviet Union as it really was.

Our two summers in Stanford were very important for us. We became acquainted with California, which amazed us because it was so different from the East Coast in climate, landscape, and lifestyle. And it gave us a chance to see many old friends, including Felix and Lore Bloch, and Isidor and Helen Rabi, who were also there for summer school, as well as Leonard Schiff and many others in the Stanford physics department. Rabi told us about his recent work, for which he later received the Nobel Prize, on the deformation of the nucleus of heavy hydrogen. That nucleus is an ellipsoid instead of a sphere. One measures this elongation using a magnitude called the quadrupole moment.

The tradition of physics summer school programs had begun at the University of Michigan in Ann Arbor in 1928, and in 1940 I was invited to teach there. The leading members of its excellent department were two Dutch physicists, Sam Goudsmit and George Uhlenbeck, the original proponents of electron spin. Because the university had always had excellent relations with Europe's physicists, all the great ones—including Pauli and Fermi—taught at the summer school in its early days. In 1940 one of the other physicists invited was Hendrik Kramers from Holland, who had been Niels Bohr's first collaborator and one of the first scientists to work at the Bohr Institute. He had a Danish wife, whom he had married after going to Copenhagen to work with Bohr in 1922, thereby adhering to the law of marriage that governed those who worked with Bohr.

My contacts with Kramers were very valuable to me. We discussed the difficulties of understanding the interaction between the electron and its electromagnetic field. This problem had haunted me and others for many years. Kramers gave me some ideas about how to perform the intricate separation of the disturbing effects of the fields in the immediate neighborhood of the electron. I followed up his suggestions only after the war, since the problems of nuclear reactions were taking up most of my time.

Two years after our arrival in the United States, Hitler invaded Poland. We were horrified to read about the successes of Hitler's armies and about the terror that was being unleashed against political opponents and Jews. For the refugee community this was a schizophrenic time. On the one hand, we felt lucky that we had escaped in time and were enjoying a successful integration into the American environment. On the other, we were deeply concerned and frightened about the events in Europe and about the fate of those who could not escape. Consequently, we were impatient with isolationists who wanted the United States to stay out of the conflict, and when the country finally entered the war after Pearl Harbor, I was eager to help the war effort in some way. We hadn't yet received our American citizenship papers, and ironically, because of the German annexation of Austria, we were considered enemy aliens. As such, we were not allowed to own a shortwave radio and were excluded from any kind of war work. It takes at least five years to become a citizen, so I had to wait until 1943 before I was considered no longer an enemy of the country that had saved me and my family from Hitler.

Some of my friends—Hans Bethe and Felix Bloch, for instance—had been here longer and were able to take part in important war work such as the development of radar at MIT, which began even before America's direct involvement in the war. This project was designed to aid the British war effort, and it helped prevent a Nazi invasion of England. When Bethe came to the radar lab in 1942, I was asked to take over his courses at Cornell University, which I enjoyed, especially since George and Anne Winter lived in Ithaca, New York.

In 1938 a fateful discovery was made in Germany. The radio-chemists Otto Hahn and Fritz Strassman found that the nucleus of the uranium atom splits into two parts under neutron bombardment. Lise Meitner and Otto Frisch, both Jewish refugees from Germany, realized that an enormous release of energy resulted from that phenomenon. They met in Sweden and correctly described the violent splitting of the uranium nucleus, which was dubbed "nuclear fission." Researchers in France, the United States, and the Soviet Union painstakingly studied the process and found that neutrons were set free during the splitting. They concluded that a chain reaction could take place in uranium: One neutron would cause one uranium nucleus to split, then the neutrons set free would hit other uranium nuclei, which would then also split, and so on. The possibility of liberating large amounts of energy for industrial use and the prospects of constructing a bomb of enormous explosive power were soon matters that occupied many physicists' thinking. The nuclear age had begun.

Niels Bohr visited America in 1939, bringing with him this exciting news. Many discussions followed, and new experiments were performed to check the results and obtain more specific information. Enrico Fermi and Leo Szilard were two of the people most involved in these endeavors. Fermi had come to America in 1938, the year during which he received the Nobel Prize for his discovery that neutrons induce nuclear reactions. Immediately upon learning about the so-called fission process of uranium, he began to investigate its different aspects in the physics laboratory of Columbia University. Later he built the first nuclear reactor in a squash court under the stands of a football field at the University of Chicago.

Leo Szilard was a unique individual. He never held a regular professorship at a university. He moved from one temporary job to the next, lived mostly in hotel rooms, and spent his life producing new and original ideas and proposing experiments to other people. He was always active at the most exciting frontiers of physics and later of biology. He was also deeply involved in politics, proposing solutions to the new problems raised by the scientific developments to political leaders in the West and the East. Born and educated in Hungary, Szilard spent most of his professional life in Berlin, England, and the United States. During Hitler's early days Szilard founded an association that helped displaced scientists survive and find jobs. He often got the necessary financial support by simply telling—not asking—rich people to give him what he needed. I was willingly drafted to be the representative of his association while in Zurich from 1934 to 1936. Szilard put his full energy into the development of technical applications of fission soon after its discovery. In the early thirties, when he had only a vague idea of the possibility of nuclear chain reactions, he even took out a patent. He was one of the first physicists to be concerned that Nazi Germany would develop an atomic bomb before the Allies did, and he actively tried to draw the attention of American physicists—particularly those who were refugees—to the impending danger.

It was Szilard who convinced his good friend Albert Einstein to write the now-famous letter, which Szilard drafted, to President Roosevelt urging the U.S. government to initiate a crash program for the development of an atomic bomb. Roosevelt received the letter in 1939, but no program was seriously started until 1943. The only action taken before that date was the formation of a so-called Uranium Committee under the direction of Lyman Briggs. It supported a few research projects, but the wartime obsession with secrecy prevented American scientists from seeing several interesting reports from the British about their progress in developing a bomb.

Obviously, so-called enemy aliens like me were not involved at all, although we would have been in a position to contribute to the work. Only Fermi, recognized as the world's greatest expert in nuclear physics, was treated as a special case and received some limited funding for his research. All of this was extremely annoying to those of us who knew a little about the possibilities. Szilard was particularly frustrated. He had tried to use his influence where he could, but with little success.

On one occasion in 1939, before the outbreak of the war, I visited Princeton and found Szilard, Wigner, and Placzek there, extremely worried about a Nazi bomb. In particular, they were concerned about Frederick Joliot and his co-workers in Paris, who were working on an experiment measuring the number of neutrons released in the fission process. They were about to publish the results of their research in a physics journal, and we thought this information could be useful to the Germans in their bomb-development project. Physicists are by nature inclined to openness and opposed to any censorship of their results. Still, the Princeton group and I felt strongly that we should arrange an agreement among the allied physicists that such results should not be published at this time.

Since one of Joliot's collaborators, Hans Halban, was a close friend of mine, I proposed sending him a cable asking him in veiled terms to refrain from publication. The cable was sent, but it did not influence Joliot. He answered that Fermi had published similar results and that he didn't see any reason for holding back the facts he had discovered. We were very disappointed. (A similar cable was sent to Patrick Blackett in Cambridge, England, but I do not remember his reaction.)

Although we were excluded from officially working on the nuclear bomb, Bethe, Placzek, and I discussed the problems involved in the project during my numerous visits to Cornell, and we tried to get some idea of what was involved in achieving a nuclear explosion. The slow and secretive way fission research was done at that time was very frustrating for us. Our fear that Hitler would develop the nuclear bomb before the Allies was a realistic one. After all, the fission process had been discovered in Germany; and although many of the physicists had left Germany because they were Jewish, the Germans still had Werner Heisenberg and other excellent scientists.

This overpowering fear of the Nazis beating us in the race for the atom bomb caused me to react in a bizarre fashion to the news that Heisenberg would visit Switzerland in spring 1942. I wrote a letter to Oppenheimer suggesting that Heisenberg be kidnapped during his Swiss visit, and I even offered my services in carrying out this scheme. Oppenheimer took it seriously enough to forward the suggestion to the military authorities. Fortunately, this ill-conceived endeavor never took place. Today I have a hard time understanding how I could have proposed such a harebrained idea, let alone considered participating in its execution.

Under constant pressure from Szilard, Wigner, and especially from Ernest O. Lawrence in Berkeley, the government finally gave its full support in 1942 for development of an atomic bomb. The Uranium

Committee under Briggs was dissolved, and a crash program was initiated under the leadership of General Leslie Groves.

Oppenheimer was selected to take charge of the scientific and technical investigation of the many processes that play a role in a nuclear explosion and in the design of the bomb. I have always admired Leslie Groves for having made this choice. In many respects Groves was a man imbued by the spirit of discipline and military organization. He was also pedantic and often narrow-minded when it came to small or unimportant issues, and politically he was very conservative. It must have been difficult for him to choose a man so different from the military type—and one who had been known as a Communist sympathizer in the past—as the leader of the most important part of the bomb project. But I have always thought that Groves sensed that Oppenheimer was the right man to lead the scientists and engineers, and he was right.

In summer 1942, Oppenheimer called a number of physicists together in Berkeley to discuss the problems that would have to be solved to produce a nuclear explosion. One of the participants, Robert Serber, wrote up what was known and could be concluded easily from available knowledge as to the probable size and force of a nuclear bomb. This report became known as "The Primer." For any scientist joining the project, "The Primer" provided a good basic idea of the project and what had to be accomplished. But at the time, I was still considered an enemy alien and was unable to participate in the meeting.

Groves and Oppenheimer had to find an isolated place where the ideas about how to construct an effective nuclear bomb could be developed and where the experiments necessary for testing the ideas could be carried out in secrecy. Oppie suggested Los Alamos, not far from his little ranch in northern New Mexico. The government bought up the boys' boarding school that stood on the site and transformed it into a laboratory for the development of the most destructive weapon ever conceived.

In order to get the brainpower necessary for the project, the government relaxed the law on exclusion of enemy aliens. Those with experience in nuclear physics could be admitted if two American scientists vouched for them. Early in 1943, Oppenheimer called me to New York for an important personal conversation. He told me about the bomb project and asked me to join him at Los Alamos. I was strongly urged to keep the location and purpose of the project secret. Of course, I accepted immediately. There was never any thought of turning this down. How could I have refused an offer to join the best people in the country in a project of such enormous importance? How could I have refused to participate in the war effort of the country that had accepted and supported us so generously? Oppenheimer and Robert Bacher were my sponsors, and I was ready to be hired officially to work on the project.

# EIGHT

⊠

# Working on the Bomb

ONE DAY EARLY IN 1943 a secretary's voice was heard calling through the halls of the physics building at the University of Rochester, "Professor Weisskopf, telephone from Santa Fe, telephone from Santa Fe." My first thought was, Everyone will know what a call from Santa Fe means; Oppie told me to keep it a secret, and now the secret will be out. When I answered the phone, I learned that I was to move to Los Alamos within a few weeks and that my family was to follow. Ellen was told where we were moving but not why we were going. She was asked to keep this information secret.

Indeed, a few days after I left for Los Alamos, a security officer visited our house in Rochester and asked Ellen where her husband had gone. Although he insisted that he had the right to know about our movements—after all, we were still technically enemy aliens—Ellen stood fast. She told him she knew, but that she had promised not to tell anyone. But as I had feared, after the phone call everyone in the physics department knew where we were going. Such have been, are, and will ever be the inconsistencies of military security.

Los Alamos was a very attractive place. It was located on a wide, rather flat expanse on top of a mesa, surrounded by green hills, ranges, and pine forests where elevations of more than 7,000 feet rose like green islands out of the dry New Mexican desert. From there one had a stunning view over the desert to the Sangre de Cristo Mountains on the other side of the valley, and over the hills and ranges of the Bandelier National Park. The country was very sparsely populated. Many of the people were Native Americans living in pueblos or in small villages with beautiful adobe homes perched around Spanish-Mexican–style churches.

Sante Fe, the largest city in the region, was a center of Native American crafts, with merchants selling beautiful authentic rugs and jewelry, as well as lots of the cheap junk that emerges when a modern civilization invades an old culture. Nevertheless, Santa Fe was (and still is) very different from the average American town. Its main hotel, La Fonda, built in imitation adobe style, was furnished with copies of furniture built in the local manner. It was a pleasure to eat in the exquisite restaurant, La Placita, located in a lovely Spanish-style courtyard. The hotel and its restaurant were to play an important role in our lives in Los Alamos. Santa Fe also had several excellent bookstores catering to an intellectual elite of wealthy people who had returned to the region because of its agreeable, dry climate and the beautiful surroundings. The influx of a scientific community was a welcome boon to the bookstore owners.

The first weeks in Los Alamos were like an exciting, romantic adventure. Since there was as yet no housing for us, we slept either in makeshift arrangements in the boarding school's houses or in the lodge, an old and rather primitive log house which had been the guest house for visiting parents of the boys attending the school. The lodge had a library, a few guest rooms, and a restaurant, which was often used as an assembly hall. (In the library I found *Seven Gothic Tales,* by the Danish writer Isak Dinesen, a book that impressed Ellen and me enormously. In spite of being Danish, Ellen was unfamiliar with Dinesen's work until we found this book in the Los Alamos library.)

In April 1943 a group of around thirty of us were gathered in this beautiful place so we could discuss how to embark on the most secret of all war projects. In the early phase of our work at Los Alamos there were many questions to be answered: Exactly how many neutrons are released with each fission of a uranium nucleus? What is the speed of these neutrons? How are the neutrons absorbed or scattered around when they penetrate different materials? How do the neutrons from one fission process produce a secondary fission when they hit another uranium nucleus? Are there other fissionable materials besides uranium, with higher yield? How does one assemble a supercritical amount of fissionable material fast enough to set up an efficient explosion? What happens during the explosion? How can one maximize the explosive power? Are there processes that absorb neutrons so that the explosive power is reduced or even prevented? It was quite a list.

Obviously, a very large laboratory with many branches working simultaneously was needed to answer these questions within a reasonably short time. Because of the urgency of the project, all necessary funds and material resources had been made available. Every day we assembled in a pleasant wood-paneled room in the lodge. During an early session Bob Serber's primer was distributed, and in subsequent meetings we discussed detailed plans. Our limited time did not permit us to follow one method until we were sure that it did not work. We had to pursue many possible avenues toward our goal simultaneously, and our work had to be coordinated with projects under way at the other laboratories that had been set up in Oak Ridge, Tennessee, in Chicago, and in Hanford, Washington. The logistics were fairly complicated.

Oak Ridge was constructing a large plant to separate uranium 235 from ordinary uranium. Most of the elements, including uranium, consist of atoms of different weight. This weight is usually given in units of the lightest of all atoms: the hydrogen atom. Uranium with the atomic weight 235 turned out to be more easily fissionable than ordinary uranium. Separating it out was not an easy task, however, because natural uranium contains only 0.7 percent of the lighter kind. In Hanford large nuclear reactors were erected to produce a new element, plutonium, with even better fission properties than uranium 235. Plutonium, an unstable radioactive element, decays rather slowly, mainly into uranium and a helium nucleus (alpha particles). A piece of uranium is itself reduced by half after 24,400 years. That is why plutonium no longer exists on earth. It must have been produced about seven billion years ago during the giant supernova that formed most of the elements of our solar system. Plutonium has since disappeared from our environment, but it can be produced artificially by nuclear reactors such as those in Hanford.

By the time we arrived, most of Los Alamos had been transformed into a huge construction site full of mud or dust, depending on the weather. The first buildings erected on the mesa were laboratories. Most important were the buildings housing the cyclotron and a linear accelerator. The cyclotron was shipped from Princeton, together with a crew of a dozen physicists. The linear accelerator—a Van de Graaff generator—came from Wisconsin. Both machines were to be used as neutron producers in order to find out how neutrons penetrate matter.

Usually it takes years of construction before laboratory buildings are ready. In Los Alamos they were built and readied for use within a few months. At the same time, residential housing was also built. Everything is possible under wartime conditions. In less than six months a large part of the rather primitive construction was finished and ready to accept cyclotrons, Van de Graaff generators, and people—including the four Weisskopfs.

The families were housed in green clapboard buildings placed between groups of pine trees in a pleasant irregular fashion. The two-story houses could hold four families with children, and smaller one-story houses were meant for childless couples. There were also large dormitories for single people. The apartments were pleasant, equipped with fireplaces and relatively large living rooms, kitchens, and baths with showers. Our house had a particularly wonderful view over pine trees toward the hills of the Bandelier range.

In our four-family house there were, besides us, two other families of physicists: John Williams and his family from Minnesota, and John Manley and his family from Chicago. David and Frances Hawkins and their child, Julie, were the fourth family to join our housing group. David was, of all things, a philosopher whose main interest was the philosophy of science.

When I heard that we had a philosopher in our midst, I insisted on having him in our house. I thought it might keep us in touch with other aspects of the human spirit in an environment of scientists, engineers, and military personnel. Indeed, it was the beginning of a lifelong close friendship, not only between the four of us but also between our children and grandchildren.

It was Oppie's good idea to include such a person in the bomb-building project. David was also a talented manager, so he was responsible for much of the day-to-day operation of the project. One might think that the duties of the philosopher would include worrying about the ethical problems of our enterprise. That concerned all of us later when the work was finished. At the outset, however, we were all convinced that our work was vital, and we did not worry much about moral issues. Later, when our work was finished, David wrote the official history of the project.

The average age of the scientists and engineers "on the hill" was late twenties, and many were starting families. Our daughter, Karen, celebrated her first birthday in Los Alamos; our son, Tom, was three years old when we moved in. Most of the children were preschoolers, and many babies were born during our stay in Los Alamos. The hospital was happily busier with childbirth than with diseases. At one point in this population explosion, General Groves asked Oppie to exhort his collaborators to reduce the high birth rate; it was overtaxing the hospital and the housing facilities and required many pediatricians. Oppie coolly reminded him this sort of counseling was not part of the job description of a scientific director. So the production of babies went on, including a second Oppenheimer baby.

The laboratory, whose highest authority was after all a general, was in a loose sense part of the military establishment. At the beginning there was even talk of giving us all military rank. Depending on their seniority, the scientists would become sergeants, lieutenants, majors, and a few perhaps even generals. But some of us feared that military discipline might not be conducive to scientific productivity. Freedom to try out ideas without explicit permission is a vital element of the creative process. After some rather acrimonious discussions with the military it was decided that we would remain civilians.

The management of Los Alamos continued to be in military hands. A detachment from the U.S. Army Corps protected us and watched over the security of the laboratory, keeping undesirable elements from penetrating the area. In addition, they took care of the maintenance of the place, providing us with heat, water, and other utilities and permitting us to use the military post's facilities. We had access to the hospital, and we shopped in the post exchange, which was well supplied with good food at a time when some things were scarce in the rest of the country. We also used the army cafeterias and restaurants.

The laboratory site was surrounded by a high fence controlled by the military. There were two entrances where we were stopped and asked for an identification badge. It was all designed to present an appearance of severe security, but it was not exactly airtight. Once one of my colleagues discovered a hole in the fence. He crawled out and returned through the official entrance, reporting both the hole and his exploit to the guard. The guards' first response was to interrogate my friend. They asked him why and how he did it, where he came from, and how he got a security clearance to work at Los Alamos. Then, in typical bureaucratic fashion, the hole was left unrepaired for several days.

As is usual in the early stages of scientific investigation, the further the work progressed, the more problems turned up. The interaction of neutrons with matter was more diverse and complicated than we had anticipated. At one point we almost concluded that the stuff would never explode because certain substances absorbed too many neutrons. We did not know at the beginning how to put the uranium or the plutonium together fast enough to avoid the explosion starting at a stage where it would not yield sufficient power. It would be impossible within this short account to mention all the different pitfalls we faced. In any case, at that point we were not certain that a bomb could be assembled, and some of us occasionally believed it would be never be possible.

The psychological reactions to these difficulties were varied. Some of us, including myself, secretly wished that the difficulties would be insurmountable. We were all aware that the bomb we were trying to develop would be such a terrible means of destruction that the world might be better off without it. And if it was impossible for anyone to develop a nuclear bomb, there wouldn't be any danger of the Nazis having one.

Then, imperceptibly, a change of attitude came over us. As we became more deeply involved in the day-to-day work of our collective task, any misgivings that we had at the start began to fade, and slowly the great aim became the overriding driving force: We had to achieve what we had set out to do.

Today I am not quite sure whether my decision to participate in this awesome—and awful—enterprise was solely based on the fear of the Nazis beating us to it. It may have been more simply an urge to participate in the important work my friends and colleagues were doing. There was certainly a feeling of pride in being part of a unique and sensational enterprise. Also, this was a chance to show the world how powerful, important, and pragmatic the esoteric science of nuclear physics could be.

I can think of no one who refused to participate in the project because of a conviction that our great science would be demeaned by serving in the manufacture of the means of death and destruction. We were all acutely aware that the whole civilized world was under attack by a force of the greatest evil. Still, the question loomed in our minds whether it was moral to use such a devastatingly destructive weapon even to defeat an enemy as undeniably evil as the Nazis. But this was an old question, one humankind had had to face many times during its history There was no clearcut answer during the days of the Los Alamos project—and there still isn't.

It is interesting to compare our attitude in the United States with the way some German scientists dealt with these problems. Three scientists in particular played an important role in Germany during the time of the Nazi regime: Werner Heisenberg, Hans Jensen, and Wolfgang Gentner.

Before the advent of the Nazis, Heisenberg was known to be a pleasant man who made friends easily. Nearly everybody who came in touch with him liked and admired him. Interested in all aspects of human culture, he was an accomplished pianist and an enthusiastic mountaineer and skier. One could have envisioned him leading a happy and fulfilled life as a renowned scientist in the midst of his many students and collaborators, a life devoted exclusively to a deeper understanding of physics and the enjoyment of art, music, literature, and the beauties of nature. But the events of history conspired to thwart such a placid existence. When the Nazis came to power, Heisenberg was forced to make certain choices that drew much criticism.

I do not believe that he was a Nazi. As a man deeply rooted in the best of German cultural life, he tried to find positive signs in the National Socialist movement. When he saw its true nature, however, he was strongly repelled by its spirit, its excesses, and its actions. What then were his choices? He could emigrate, he could actively participate in the underground movement, or he could retire from public life and live as decently as possible with his family and his work. The last possibility existed for many other Germans but not for him—he was much too influential and prominent. Participation in the underground movement would have required an amount of heroism that one cannot ask from any person. Heisenberg was the head of a family that included five children. He was not the heroic type, but rather a careful and prudent man. Active opposition to the Nazis could not have been his way.

He did not choose to emigrate. Many of his friends living abroad strongly encouraged him to do so and would have made it feasible. His life and that of his family would have been much easier. But he thought it would be a dishonest choice. He wanted to stay in Germany and help create "islands of decency" where, within small groups, some of the country's cultural achievements might be saved to serve as the basis for a new cultural life when the Nazi reign of terror was over. Max Planck had told him, "In the ghastly situation in which Germany now finds itself, no one can act entirely decently." Heisenberg would learn that this was true.

Obviously, by remaining in Germany, he soon became involved in the efforts of the Nazi regime to exploit nuclear energy, which began as soon as the possibility of a bomb became evident. The scientists on the Allied side and those on the German side, specifically those who were more or less opposed to the Hitler regime, faced vastly different situations. The Allied scientists were eager to work on the bomb because they trusted the policies of their leaders, Churchill and Roosevelt, and wanted to prevent the other side from developing the bomb. Those German scientists who, like Heisenberg, were not in favor of the Nazis would have been in a terrible situation if they had been called on to devise this awesome weapon.

Heisenberg and his colleagues were asked to evaluate the time and effort necessary to construct a nuclear bomb. They estimated that it would take several years. On the basis of that information, the German Army Ordinance decided late in 1942 against an all-out effort to construct the bomb. They did not want to start a new weapons project that would not produce results for several years. They suggested reducing the nuclear effort to research and development on nuclear power reactors. In June 1942, Heisenberg discussed this issue with Admiral Speer of the German navy. Both agreed that it would make no sense to go ahead with such a big enterprise under war conditions. Heisenberg reported his findings to the Nazi chiefs later in 1942. The decision not to make the bomb saved Heisenberg and his colleagues from many problems. Nazi authorities would never have let them work without stringent supervision; unavoidable setbacks would have been interpreted as treason, and certainly many people would have been harassed, if not persecuted. Furthermore, Heisenberg and his group would have faced the dilemma created by the call of their conscience versus their sense of duty.

In 1942, Heisenberg went to Copenhagen to talk to Niels Bohr, his old friend, mentor, and father figure, about the problems that the development of nuclear explosives had created for humankind and, in particular, for the community of scientists. It turned out to be an unfortunate and abortive visit. Heisenberg did not comprehend the deep hatred and terrible desperation that permeated Denmark, a nation victimized by the Nazis. He expressed himself vaguely, fearing that any direct statement about Germany's nuclear effort, or any doubt of a German victory, would put him and his family in mortal danger. Under these conditions, it was difficult for Bohr to see Heisenberg as a disciple and friend and not as a representative of the oppressor. For his part, Heisenberg could have been expected to trust Bohr. He should have assumed that his old friend would have taken every precaution to prevent their being overheard and would never have given him away. An open exchange could thus have been useful, but it did not take place. Consequently, a great friendship was shattered, and a creative human bond was severed. Certainly, an end to a friendship cannot be compared to the mass murders and the effects of mass bombings, but it is a symbol for the tragedies of war.

In times of oppression and persecution, general principles can be less important than personal actions. We know that Heisenberg did what he could to protect the Bohr Institute during the German occupation, and he saved many people's lives. He never made much of this and did not mention these efforts in his writings, but there is clear published evidence of his actions. Such acts carry more weight than any statements. I saw Heisenberg after the war, and he was completely changed from the man I had known. Before the war he had always struck me as an innocent Boy Scout type, free of worries, youthful and enthusiastic. But when I saw him again, even his complexion had changed, and this was not due only to age. He visibly carried a load. I could not help thinking of Oscar Wilde's *The Picture of Dorian Gray* when I saw the imprint of those tragic years on his face.

In stark contrast to Heisenberg, J. Hans D. Jensen, the German physicist who later shared the Nobel Prize with Maria Göppert Mayer, went to visit Bohr at about the same time and told him outright that the Germans were not planning to build the bomb. Unfortunately, Jensen did not belong to the Copenhagen clan. Bohr did not know him well enough and thought he was ordered by the Nazis to mislead the Allies. In fact, Jensen was connected to the underground and risked his life by speaking with Bohr, who discounted everything Jensen said.

Wolfgang Gentner, who did important work in several fields of physics, also remained in Germany for reasons different from Heisenberg's. His character was very different as well; in many ways he was Heisenberg's antagonist. He stayed to fight the regime in every way he could, becoming actively involved with the resistance movement, and survived the Nazis only because he was never found out.

During the Nazi occupation of France Gentner was sent to Paris to take over the French Institute of Nuclear Physics, which had been directed by Frederick Joliot. Joliot's leftist sympathies were well known. It was widely assumed that he belonged to the Communist party and participated actively in the French resistance movement. In fact, under Gentner's protection, Joliot was making bombs and weapons for the underground resistance in the basement of the institute. The discovery of this dangerous work would surely have resulted in Gentner's execution. After the war he was made an honorary citizen of France for his heroic stand. He managed to avoid any discussion and to distance himself from any research connected to the nuclear bomb, although he was experienced in nuclear physics. His example demonstrates that a well-known physicist could stay away from work on the bomb if he was deeply convinced that an atomic bomb in Hitler's hands would lead to ultimate catastrophe.

These three examples are by no means characteristic of the majority of German physicists. As in any country, and in particular one under an oppressive regime, most of the scientists followed the orders of their superiors and did not openly question the wisdom of their decisions.

The type of work in which we were engaged and the urgency of its rapid progress made it necessary for us to proceed in a manner unfamiliar to most of us. We were accustomed to working alone or in very small groups. But now extensive teamwork became essential, even among the theoretical physicists. Luckily, we had not yet developed the rigid work habits and entrenched ideas that might have hampered older scientists. Groups of a dozen or so scientists had to collaborate closely on one specific, well-defined subject. In each of these teams there was always one person or a handful who played a more dominant role. They were the ones who had most of the ideas, who were more inventive about getting around obstacles, who pushed on when the problems seemed insurmountable, and who exerted leadership.

Collaboration and cooperation lent a definite character to our project and added considerably to the challenge and excitement of our work. In the years that followed Los Alamos, this became the way fundamental non–applied physics research was often carried out. In the postwar years the complex, large new instruments of research, such as the giant accelerators, and the elaborate data handling often made large teams a necessity. Our experience helped facilitate that transition when it came.

Oppenheimer, an unusually inspiring leader, had an extraordinary talent for grasping the essential points of a problem, even in fields far removed from his special training. His ability to be ready with the answer before one had finished formulating the question helped him to be aware of everything interesting that happened on the hill. He was able to suppress his personal sympathies and antipathies, and he controlled his well-known impatience with people who didn't grasp concepts as quickly as he did.

Whenever an important technical discussion about a problem took place, Oppie appeared out of nowhere and helped solve it. Whenever an experiment reached a decisive stage, Oppie was present, even if it was three o'clock in the morning. Everybody felt his encouragement, his readiness to help, and his interest in any personal achievement. (In this respect he became my role model, when many years later I was appointed director of the large scientific-technical enterprise at CERN, in Geneva.)

We all agreed that the work we were engaged in was fascinating. The temperatures expected at the center of the explosion were almost a hundred million degrees, about ten times higher than those in the center of the sun. The pressures were many thousand times larger than any pressure ever produced in a laboratory. How could we predict what would happen, what the speed of expansion would be, or how matter would behave under those circumstances? We could not get the answer by duplicating those conditions in the laboratory. All we had was imaginative guesswork. By extrapolating our knowledge to an extreme degree, we could get some idea of what would happen. In addition, we tried to get the advice of the most knowledgeable specialists in shock waves and in high-pressure and high-temperature phenomena.

Sir Geoffrey Taylor came from London to talk to us about explosions, shock waves, and instabilities in the motion of matter. But no expert had ever had to deal with such extremes of pressure and temperature. It was the theoretical physicists' hour in the spotlight. We had to calculate the behavior of matter under the most unusual conditions. The available knowledge of the structure of materials is a reliable guide for the prediction of material properties only if the conditions are not too far out of the ordinary.

This is why the theoretical division was so important at Los Alamos. We tried to get the best people available. Hans Bethe was the division leader, and we had Enrico Fermi, Robert Christy from the California Institute of Technology, Edward Teller, Philip Morrison from Berkeley, and later Rudolf Peierls from England. My old friend

Placzek came, and Robert Marshak joined us from Montreal, where he was working on a British–Canadian project.

Richard Feynman came to Los Alamos as a young man. Later he became one of the foremost theoretical physicists of our age. In spite of his youth, his quick and creative mind was already evident; he proved extremely useful in solving difficult problems. He also added to our group because of his charm, playfulness, and personal warmth. Extremely witty, he was a master of the practical joke. For instance, he could program our mechanical calculating machines to produce a rhythmic clatter that mimicked popular songs of the day. He was also expert at opening locked safes, an activity the military staff did not appreciate. To our children, who adored him, he was the funniest man in the world and their favorite adult playmate.

Feynman's subsequent contributions to science were enormous. Modern field theory would be unthinkable without his decisive work. He conceived seminal ideas in almost all fields of physics. The only theorist I can think of who was comparable to Feynman was Lev Landau. What impressed me about Feynman was his intuitive understanding of physics as well as his expertise in mathematical formalism. Many times after the war I went to Pasadena, where he was a professor at the California Institute of Technology, and asked for his help in understanding this or that physics problem. He was always ready to discuss anything with me and always managed to use just the right language, leaving out mathematical complications. He knew what appealed to me. Unfortunately, like Landau, he died too early. (Feynman died in 1988 at the age of seventy.) It will be a long time before we will have such a brilliant, original mind in our community. In addition to admiring him for his scientific ability, I was deeply fond of him and miss him.

At Los Alamos difficulties arose with Edward Teller, who adamantly wanted to develop the fusion bomb (H-bomb) simultaneously with the fission bomb we were all working on. Oppenheimer finally had to let him have a special division with a few collaborators to work on the "superbomb" as we called the H-bomb at the time. I had a personal conflict with Teller when Bethe asked me to become deputy division leader. Edward maintained that he was the better physicist and should have been given the job. I didn't try to deny the allegation but pointed out to him that Hans probably had chosen me because I was better in dealing with people. Teller's political conservatism, which continued to grow, prevented any reestablishment of our friendship in later years.

We expected to receive the first shipment of plutonium from Hanford sometime in 1944. It was to be a little grain, perhaps a millimeter across. Up until then, the quantities of plutonium produced by cyclotrons and other means were so small that the physical properties of the substance had not been established. We were certain of only one fact: the plutonium atom has ninety-four electrons. That number does in fact determine most of the properties of the atom. So powerful is our knowledge of atomic structure on the basis of quantum mechanics, that knowing the atomic number should make it possible in principle to figure out how the atoms combine to form the material we call plutonium and to deduce some of its chemical properties. In this sense, quantity defines quality, but this is true in principle only. Actually, such calculations are extremely complicated and cannot be performed with any degree of reliability. Atomic physics predicted that plutonium should be a metal, with a specific weight of about eighteen grams per cubic centimeter, of brownish color, with this elasticity and that electric and thermal conductivity, and so on.

When the plutonium came, I held on the palm of my hand the first little grain any of us had ever seen. (I should not have done it, I suppose, because of its radioactivity, but it was such a tiny quantity that it didn't have any detrimental effect.) It was indeed a heavy, brownish metal, and it had the properties that my colleagues had deduced from the number ninety-four. This gave us all some further confidence in the power of scientific insight.

Part of my assignment was to study and predict what would happen when neutrons hit the nuclei of atoms in matter. A neutron can penetrate solid matter as far as a few inches before it hits a nucleus, because nuclei are so much smaller than atoms. When a neutron hits a nucleus, it may be scattered or absorbed, or when it is a uranium or plutonium nucleus, the collision may induce fission. These effects depend critically upon the speed of the neutron. It was impossible to measure all these effects because it was difficult at that time to produce neutron beams of any specified speed at will. Much had to be deduced from general knowledge, guessed at, or extrapolated from other measurements.

Frequently I had to use my intuition. Intuition in science consists of half-conscious knowledge—a certain feeling of how things work even if no exact information is available—and being right depends on a certain amount of luck. On one occasion the measured amount of uranium fission produced by very slow neutrons seemed to be much lower than expected. Of course, measurements are measurements, so this finding supposedly had to be right. Nonetheless, based purely on my intuition, I predicted that at slightly higher speeds the fission yield would increase about threefold and reach the expected value. I felt the low yield for very slow neutrons was an unexpected fluctuation. Had that yield been low over larger stretches of neutron velocities, the amount of fissionable material necessary for the bomb would have been greater.

At first my colleagues did not believe me. When it turned out many months later that the yield indeed went up to the predicted value, my office was dubbed the seat of "the oracle," or less reverently, "the cave of hot air." I was constantly asked to predict neutron effects. Mostly I was right; sometimes nature knew better.

As deputy chairman of the theoretical division, I had to become acquainted with all the other tasks of our group. Nuclear physics problems were not our only focus. All fields of physics had to be explored if we were to understand and predict what would happen during the explosion. We were involved in a unique learning process and had the opportunity to get acquainted with the newest developments in materials science, hydrodynamics, thermodynamics, and other areas of physics that we might ordinarily not have studied. We were forced to go way beyond the limits of current knowledge in order to get some idea of the processes that would be taking place during the explosion.

It was important for us to know how strong the explosion would be and how matter would be hurled apart. And we were particularly concerned with how much of the radiation would escape and spread in the air. None of these questions could be answered by experimentation because the only way to duplicate the circumstance would have been to explode a nuclear bomb. We had to know these effects beforehand and had to rely on theoretical studies for most of our knowledge.

Later we found that we had been wrong in quite a few instances, but never very seriously wrong. For example, we underestimated the amount of so-called gamma radiation (this is a very high frequency light emitted by atomic nuclei when hit by energetic particles) that would emerge from the exploding body because it was too difficult to determine the absorption of radiation by matter at extreme pressures and temperatures. In this case it would have been better if our figures had been right; fewer people would have died of radiation sickness in Hiroshima and Nagasaki.

The fascination of our work took hold of our minds. Never before had my colleagues and I lived through a period of so much learning, of so many new insights into the structure of matter in all its manifestations. For us it was a heroic period, but we also had to consider the terrible destructive potential of our work. Studies were based on experience with ordinary bombs, and it was difficult to predict what would happen when the force was at least a thousand times greater.

We tried to determine the degree of destruction, the number of victims if the bomb exploded over a city, and the potential for radioactive damage to humans, animals, and soil. All this required painstaking research in our laboratories and at our desks. Under the circumstances we were unable to confront the moral issues of our work even though we recognized them. There is no denying that constant discussions about the nature of the damage caused by fire and radiation sickness, and about the millions of deaths led to a growing numbing toward those terrible consequences. Sometimes, perhaps in the middle of the night, some of us became suddenly aware of the horror that could be caused by our work, but we were also convinced that what we were involved in was important for saving the world from the forces of Nazism, and that kept us going. I have often wondered what our attitudes would have been had we known that there was no seriously competitive Nazi effort toward a bomb.

As time went on, it became increasingly obvious that plutonium was the best material for a bomb. It was easier to produce than uranium-235, which had to be separated out from normal uranium.

Plutonium also had a larger potential for fission when hit by neutrons, and it released more neutrons in the fission process. Because of these properties, the problem of assembling an amount of plutonium that would lead to an explosive chain reaction was more difficult than it would have been with uranium. It had to be done very quickly. The "gun" method of shooting two subcritical chunks at each other was reasonable for uranium-235, but not for plutonium. The only promising way of achieving the reaction was by quickly compressing a subcritical amount of plutonium to about twice its density, making it supercritical for a nuclear explosion, since at higher density the neutrons have a shorter path to another plutonium nucleus.

We achieved this by embedding a subcritical sphere of plutonium in a much larger sphere of ordinary high explosive, and detonating the latter with detonators placed on the outside surface. The resulting explosive wave traveled toward the inside and compressed the sphere of plutonium in a process called "implosion." This idea was proposed by an American physicist, Seth Neddermeyer. Experts on explosives were needed, and many experiments had to be conducted to find out whether and how such an implosion could be accomplished symmetrically so that the plutonium would be compressed, not just deformed. The valleys around Los Alamos reverberated with the noise of the various test implosions, for which we used explosive lenses that concentrated the explosion toward the interior.

Another difficult question we faced was how to start the chain reaction once the plutonium ball was fully compressed. A strong source of neutrons had to be placed in the center of the ball, so neutrons would be available when the explosion was supposed to start, but not before. Solving these problems required many detailed studies. Such implosions had never before been performed, nor had a neutron source ever been used in this way. The laboratory grew larger with every new problem. As new special projects were initiated, more people were needed, and the town of Los Alamos grew accordingly. It also became more international in character as physicists arrived from England and Scandinavia, joining the large group of outstanding American and Canadian scientists and engineers.

One day I was asked to drive to the nearest railroad station, in Lamy, New Mexico, to meet three British scientists and bring them to Los Alamos. I was not told who they were, and I hoped that my slight Austrian accent would not offend them. When the train arrived—the famous silver Superchief that ran from Chicago to Los Angeles—three figures came down the steps. I recognized them immediately. These "Englishmen" were Rudolf Peierls and Francis Simon, both originally from Germany, and Egon Bretscher, born in Switzerland.

Most of us were working harder than we ever had, but we enjoyed a certain amount of social life, too. There was more to life in Los Alamos than just the work on the project and the discussions about its ramifications. From time to time we went to Santa Fe to browse in the fabulous bookstores or eat a delicious meal under the starry sky in La Placita. We were an unusual community, an international crowd of extremely creative people, and even our informal social gatherings were extraordinarily stimulating and interesting. The main topic of discussion remained the progress of the war, but we also spent a lot of time talking about music, theater, and sports.

It was an ideal place to bring up small children. There was not much traffic on the streets, and the security fence that surrounded the facility kept them from getting lost. They had the company of lots of children of the same age, and there were wonderful opportunities for hiking, skiing, and playing. The school was excellent, and the teachers were mostly the highly trained and experienced wives of the scientists.

In addition, the Jemez mountains in the west and the Sangre de Cristo range in the east provided wonderful trails for Sunday outings. A few miles east of Los Alamos was the Valle Grande, an old volcanic crater that had become an immense circular expanse of open grassland surrounded by mountains. Nearby on Sawyer's Hill a group of us found an excellent skiing slope where we set up a primitive ski lift. After several years of use the lift needed a new rope. It was just before the bomb test at the site Oppenheimer had dubbed "Trinity." During his last checkup of the tower from which the bomb was to be exploded, George Kistiakowsky, never at a loss for ideas, noticed the rope that had been used to hoist the bomb up to the top. The rope had served its purpose and would have been evaporated by the explosion. George rescued it from the tower, and it carried us up the slope for many months thereafter.

We also explored the towns of the area. I will never forget our visits to the Indian and Spanish villages. In the adobe churches we discovered masterpieces of sculpture and painting combining Indian artistry with Spanish-Catholic tradition. I am still haunted by one wooden head of Christ, his penetrating eyes and his face mourning the fate of humankind. There was also a wooden chariot steered by Death in the person of an emaciated figure staring into infinite space and time. He aimed arrows at the beholder and touched my deepest roots of fear and awe.

Among the more frivolous social events of our life on the mesa were the famous dormitory evenings. At these dancing and costume parties we exhibited our various talents for entertainment, which ranged from classical music performances to elaborate acrobatic exhibitions. One time Otto Frisch and I came as a pair of dictators: he was Hitler and I was Stalin. I borrowed Hans Staub's official Swiss military cap and pasted a Soviet star on it. In Zurich this would have provoked an international incident.

By June 1943, Los Alamos had developed into a community with its own social problems, and a town council was formed. The council members and its chairman were elected by the people living on the mesa. The council was purely advisory to the military administration and had no power except in minor matters. Its main function was to serve as an open platform for discussions of community problems and as a representative body for negotiating with the military administrator of Los Alamos. It was a clearinghouse for discussions about the problems of daily life on the mesa and an opportunity to let off steam about some military restrictions that seemed senseless to us. The council was elected twice a year, and in spite of the military's distrust toward a democratic body within an authoritarian regime, we succeeded in having a representative of the military administration in attendance at each of our meetings.

As a recent emigrant from Europe, I considered it a special honor, a testimony of trust and confidence from my fellow residents, to be elected to the council for three terms and to be made chairman for one period. Of course, there were a number a foreigners like me at

Los Alamos, but still my election was an impressive example of how immigrants were (and are) so often accepted as equals in the United States, in stark contrast to Europeans' many prejudices against foreigners.

The town council sessions were an interesting counterpart to our scientific and technical tasks. Many trivial and not-so-trivial housekeeping problems were discussed and negotiated. Playgrounds for children, construction of safe sidewalks along the main thoroughfares, the organization of the post exchange (or PX), shortages in the milk supply, the recruiting of household help, problems concerning the organization of the schools, the allocation of rent payments for the apartments, and many other mundane things of that sort all came under discussion. There was also the problem of social distinctions. The machinists and other supporting staff lived in trailers and more primitive housing that did not get the attention and repairs given to our living quarters. By bringing the matter up at the council meetings we were able to improve that situation a bit.

The population as a whole participated actively in these discussions, which were almost always conducted in a spirit that focused on the success of the project. Certainly, some frictions and disappointments persisted, especially when proposals had to be negotiated with the commander or with representatives of the laboratories. Some of the military considered the council an unnecessary nuisance. But on the whole, it was a useful institution that helped reduce the frictions between the population and the administrators.

At one memorable session two issues were on the agenda. One was the allegation that the PX was charging its customers above the legal ceiling prices imposed on all commodities during the war. The other was a complaint that prostitution had established a beachhead on the mesa. In testifying about the second complaint, an inhabitant of the women's dormitory maintained that she had heard unambiguous noises attesting to the activity in neighboring rooms. In regard to the first complaint, one male member of the community, who was adept at seeing the connections between things, stated that he personally had nothing against prostitution as long as the ceiling prices were observed. The committee appointed to investigate the matter of prostitution reported at the next session: "Fifty percent imagination and fifty percent frustration." The matter was dropped.

On the whole, our relations with the military were friendly and collaborative. But the military mind was so different from the scientific that occasional frictions were unavoidable, and some of these incidents were hilarious. Every Wednesday the whole scientific community assembled in the movie house, the largest hall available, for a colloquium about the progress of different parts of our work. At first Oppenheimer had a hard time convincing General Groves that these discussions were necessary to inform the scientists about all aspects of the work and promote cross-fertilization of ideas. Groves would have preferred each scientist to know only about the work in which he was directly involved. Finally Groves gave in, but he insisted that the police check the ground under the hall for explosives. He feared that some saboteur might try to eliminate the people who were working on the bomb. So two soldiers crept under the hall fifteen minutes before the colloquium to search.

One day a physicist who liked practical jokes bet Major Peer de Silva—who was responsible for the police—fifty cents that he could detonate a firecracker under the hall at the next colloquium. De Silva didn't like the idea at all, and he ordered fifty men to search beneath the hall for two hours before the next colloquium. Of course, our friend hadn't put any firecrackers there. But at a subsequent meeting, while the major was present, he announced triumphantly, "I now have incontestable proof that Major de Silva values fifty cents more than the cream of American physics." The major was not amused, but the others were.

In fall 1944 we were told that two celebrated European physicists were expected to join us. None of us had ever heard of Nicholas and Jim Baker. When they appeared, they turned out to be Niels Bohr and his son Aage, using code names for security reasons. Bohr and his family had been evacuated to Sweden during the Nazi occupation of Denmark. When the Nazis first occupied Denmark, they considered the country a model protectorate and left the Jews alone. But when an active underground movement began to attack the occupiers, the Nazis gave the order to deport Danish Jews to concentration camps.

Fortunately, a German embassy officer, Georg Duckwitz, informed the Danes of this event a week before it was officially announced. Helmut von Moltke, who was executed by Hitler in July 1944 because of his involvement in an attempted putsch, also warned the Danish underground of the deportation of the Jewish population. Under cover of night the Danish Jews were assembled secretly at different points along the eastern coast of the country. Danish fishing boats transported them to prearranged points in neutral Sweden, where they were welcomed and invited to stay. Nearly all of Denmark's Jews were saved in this way. It was the only country under Nazi occupation that was able to accomplish this. After these events the king wore a yellow arm band with the Jewish Star of David on it whenever he rode through the city on horseback. These heroic facts are still not sufficiently recognized. This knowledge added to my admiration of the Danish people, who had given me both a great teacher and an ideal companion for life. After the war, the Danish leader, Hans Hedtoft said that Duckwitz was one of the men who made him believe in the possibility of a new Germany. Duckwitz became Germany's first ambassador to Denmark after the war. He was made an honorary Danish citizen and spent the rest of his life in Denmark.

Shortly after Bohr was evacuated, he and his family left Sweden and went to England. Then, with his physicist son, he came to America, where he was invited to join the project in Los Alamos. Bohr's arrival was an important event for all of us. Only Einstein inspired the same kind of awe and veneration that the scientific community felt for Bohr.

His contributions to our work at Los Alamos raised the intellectual level in more than one way. He participated actively in our work on scientific and technical problems, providing a number of important ideas. But he also raised our consciousness in another fashion. We were well aware that we were working on a project that led to some of the most difficult questions a scientist could face. Through physics, our beloved science, we had been pushed into the most cruel part of reality, and we had to live through it. As I have said, most of us were young and somewhat inexperienced in human affairs.

Bohr immediately involved us in private discussions on the significance of what we were doing. He did not shy away from the most problematic aspect of our scientific activity—the application of science to death and destruction. He faced it squarely as a necessity, but at the same time his idealism, foresight, and hope for peace helped us see the sense in all these terrible things. He inspired many of us engaged in the work of war to think about the future and to prepare our minds for the task of peace that lay ahead.

Bohr believed that in spite of the inevitable carnage and devastation a positive future lay ahead for this world transformed by scientific knowledge. The superweapon under construction would perhaps make world wars obsolete. (Now, forty-five years later, it seems that he may have been right.) Those of us who had worked with him in the past had learned from him that every great and deep difficulty bears within itself its own solution. Bohr's conversations with his old and new friends raised some of the questions we had suppressed under the pressure of the work. He started our soul-searching and inspired us to begin to hold regular informal discussions about these issues.

Bohr's basic instinct was to face the essential dilemmas. He had the gift of seeing all sides of a problem, and he could recognize the positive possibilities in even the most destructive of all weapons. His thinking brought a certain spirit of hope to our discussions because he so often pointed out that the existence of such devastating weapons might demonstrate that wars are useless and suicidal. He foresaw the terrifying potential for a future nuclear arms race, but he also recognized in this impending threat a unique opportunity for changing the course of history.

Bohr hoped that the temporary superiority of the United States in nuclear weapons could be used to persuade others that international control of nuclear weapons and nuclear power production would be advantageous to all. The best means of reaching this aim, he thought, would be early discussions between the nations, particularly the Soviet Union, before the weapon was actually developed and put to use. Although the Soviets were our allies in the war against the Nazis, it was expected even then that the USSR would emerge in the post-war world as the chief competitor and opponent of the United States. Bohr was convinced that the scientific information involved in making the bomb could not be kept secret and urged a policy of free exchange of knowledge as soon as possible. He always thought that any remaining national secrecy after the war would only promote further tension.

I recall meeting with Willy Higgenbotham, Robert R. Wilson, Hans Bethe, David Hawkins, Philip Morrison, William Woodward, and others on various occasions during that time. The European war seemed to be turning in favor of the Allies at the end of 1944, and that, as well as the steady progress of our project, caused us to think more about the future of the world after the war. We hoped that in peacetime the Allies would share their secrets and establish an international administration of all military and civilian nuclear power applications. We were convinced that nuclear weapons should not be nationally controlled. Bohr had created in many of us the hope that the superweapons would make future wars impossible.

We felt optimistic, too, about the possibilities of using nuclear energy for power production and medical purposes in peacetime. The dangers of radioactive pollution resulting from reactor accidents or the many other problems of the peacetime uses of atomic energy were not things we thought about. We predicted that this new technology would present a great opportunity for truly international cooperation, both for the purpose of preventing international conflicts and for developing beneficial applications of atomic energy. The varied possibilities inherent in atomic energy would be catalysts, we felt, for the idea of a supranational science and technology after the war ended.

Only part of Bohr's time was spent in Los Alamos. He frequently traveled to Washington and London and used his reputation and his close friendships with many influential people to draw their attention to the pressing questions we faced. It was very difficult for Bohr to talk to the leading people about these matters without violating the project's secrecy, but he persuaded a number of important leaders that his approach might be the only way of saving the world from nuclear annihilation. Even Roosevelt seemed to have been impressed by his arguments. However, Bohr was unable to convince Churchill, and their encounter was a tragic failure. Bohr's soft and discursive manner was such a contrast to Churchill's bold approach that the two men did not communicate at all. Churchill even suspected him of being in league with the Soviets.

In hindsight it appears improbable that any international cooperation with Stalin on nuclear matters could have succeeded. Furthermore, nationalistic factions in America and England argued against internationalization under the mistaken belief that the Western monopoly would not be broken for many years. Bohr recognized by 1950 that his attempts to arrive at an international understanding about nuclear matters had come to nought. He then wrote an open letter to the United Nations in which he predicted that the lack of such cooperation would engender an ever-escalating nuclear arms race, plus increased tensions and confrontations between East and West. His predictions turned out to be tragically correct. However, today it looks as if Bohr's vision of international cooperation and an end to the arms race may become a reality. The kind of world he had hoped for seems to have become a possibility more quickly than any of us could have imagined. It shows once again Bohr's profound grasp of the historic changes that were inevitable.

In March 1945 about forty of us got together to examine the role of the atomic bomb in world politics and of atomic energy production, as well as the status of the scientific community after the war. I was among those appointed to a committee that was to continue these deliberations. Because of the tremendous pressure under which we were working in preparation for the first bomb test, scheduled for July 1945, there were no further meetings of the committee until after the war.

After the middle of 1944 our work on the bomb intensified considerably. Many new people came to the hill to carry out the numerous detailed tasks necessary for the completion of the work. New imagined or actual difficulties turned up, some so serious that as late as the beginning of 1945 we were not sure whether the bomb would explode with a reasonably high efficiency. For example, there was a possibility that uncontrolled neutrons would produce a premature explosion before the plutonium was fully compressed, resulting in a relatively small yield.

These were some of the problems we were dealing with when Hitler was defeated in April 1945. The children of our neighborhood, none over five years old, went into the street and held a victory parade. In celebration they hit pots and pans with wooden spoons, making a joyful sound. Small as they were, they instinctively felt that something great had happened.

With Hitler defeated there was no danger that the Nazis would develop their own bomb, but this did not rise to the surface of our consciousness. By then we were too involved in the work, too deeply interested in its progress, and too dedicated to overcoming its many difficulties. We were committed to the project not only for pragmatic reasons but also because of the purely scientific search for answers. No matter what happened in Europe, we knew our work had to be completed. Only two people, Volney C. Wilson and Joe Rotblat, left the project because of the end of the Nazi threat. Joe left also because he wanted to go to Poland to try to find his family. Tragically, he learned that his wife and her parents had been killed by the Nazis, but he did find some of his own family, who had survived the Holocaust.

The reaction most of us had was both interesting and somewhat depressing. It showed how deeply we scientists got attached to the task that was set before us and to the solution of the remaining technical problems. Of course, there was also a pressing political reason for accomplishing the task. Ending the Japanese war abruptly by whatever means could save millions of American and Japanese lives that would surely be lost in a prolonged war or in the expected invasion of Japan by the Americans. Still, in retrospect, I have often been disappointed that, at the time, the thought of quitting did not even cross my mind.

Oppenheimer, of course, was very interested in all of these questions of the effect of the bomb on the political future of the world, and he had many discussions with Bohr. Like Bohr, he stayed away from gatherings at which the political problems were discussed. He thought we should not get involved in questions about the use of the bomb and of postwar nuclear politics. He believed such decisions should be left to the responsible authorities in Washington, and he expressed his own confidence in their judgment. Oppie was our hero and mentor. He so strongly influenced our thinking that we did not discuss alternatives for the use of the bomb other than the destruction of Japanese cities. Oppenheimer believed that only such a terrible lesson in the reality of the new weapon's power would make the world aware of the fundamental change in the nature of warfare that was born with the bomb.

Discussion groups like the one we participated in had been formed in Chicago and Oak Ridge, but before the end of the war no communication was permitted between us. Indeed, most of us, myself included, had no idea that in June 1945 a group under the leadership of Leo Szilard and James Franck had submitted a written proposal to the secretary of war urging the United States not to use the bomb over an inhabited location. Had I known about this proposal, I certainly would have joined the group that framed it.

The test explosion was to take place on 16 July 1945. Preparations had been made several months before at the Trinity site. Located in the middle of a desert north of Alamogordo, about 200 miles southwest of Los Alamos, the site was surrounded by low hills. Apart from the McDonald farm with its solitary farmhouse, this part of the desert—a dry, uninviting stretch of sand—was uninhabited. Plant life grew sparsely near the few waterholes. Purchased by the government, the McDonald farmland had one water hole that served as our primitive swimming pool when the temperature climbed to over 100 degrees, as it did on most days. The desert was appropriately called Jornada del Muerte, Journey of Death, to commemorate a Spanish army detachment that had perished there from the intense heat.

A crew of several hundred Los Alamos physicists and technicians spent some months at the farmhouse under the most primitive conditions, constructing the tower on which the test bomb was to be exploded and installing the various devices we would use to measure the effects of the explosion at different distances. The power and duration of the shock wave, the heat radiation intensity, the radioactive radiation (particles and gamma rays), and many other effects were to be registered to determine whether the explosion had gone according to plan and, if it didn't, to find out what caused the problem.

Since I had been in charge of calculating the various effects of the bomb, I was one of the very few theorists sent out to the Trinity site. In my battered jeep I drove from one test station to the other to talk to the crews about the expected radiation intensities and shock pressures. According to our calculations, the dangerous radiation would diminish to a harmless level about five kilometers from the center of the explosion. Several jokers suggested that I should be put there in an iron cage. We underestimated the radiation intensity because we did not know how fast the outer layers of the bomb would expand under pressures that were over a thousand times larger than anything investigated before. I would have perished had I been there.

Life at Trinity was rough, with terrible heat and dust and austere accommodations. Whenever my jeep went over a bump in the road, a cloud of dust rose up through a huge hole in the floorboards. For relaxation we swam in the muddy McDonald watering hole, watched an occasional movie projected on a makeshift outdoor screen, and admired the beautiful clear nights under a canopy of glittering stars, including the most intense Milky Way I have ever seen.

The excitement and tension were enormous. We were about to release cosmic energies and temperatures never before experienced by humans. Would it work? Some speculated that the explosion might ignite the atmosphere and spread over the entire globe. We had made many calculations and experiments that assured us with overwhelming certainty that such ignition would not and could not take place. But one of our colleagues was disturbed by an unfortunate jocular remark by Enrico Fermi during a ride the two of them took from Los Alamos to the Trinity site. Fermi said, "It would be a miracle if the atmosphere were ignited. I reckon the chance of a miracle to be about 10 percent." We all laughed at the vast exaggeration, but our colleague did not take it as a joke. Shortly before the explosion took place, he suffered a nervous breakdown.

We devoted much thought to the possible effects of the radioactive cloud that would follow the explosion. Under normal weather conditions it would spread eastward, where there was no human habitation for several hundred miles, and rise toward the upper atmosphere, where it would no longer be dangerous. But unusual weather with a downdraft in the wrong direction could not be completely ruled out.

An Indian village, Carrizozo, located about sixty miles away, would have been exposed to dangerous radiation levels if the wind were to shift. Measuring stations were placed all over the region to monitor the spread of radioactivity after the blast. For the sake of security we could not warn the village's residents. Hidden away, out of sight of the population, sixty trucks stood ready to evacuate them if needed. We could only imagine the turmoil in Carrizozo if it became necessary to remove all its inhabitants in the early morning hours.

The shot was planned for midnight. All through the months of preparation the weather had been hot, the sun fierce and unmerciful in a blue sky. In the early morning of the test day clouds appeared, and some raindrops fell during the evening. The meteorologists were in despair. Postponement posed difficulties. President Harry Truman was at Potsdam conferring with Churchill and Stalin. He expected notification of our success. But because of the weather, the shot was rescheduled for five o'clock the next morning. This postponement had a number of ludicrous consequences. For example, the inter-communication system of the test site was tuned to a wavelength that was also used by a local radio station. No interference had been expected from this station, since it was off during the night. At five in the morning, however, the station began its program of light classical music. Thus, the countdown for this test of the most powerful weapon of war yet dreamed up was accompanied by the lyrical sounds of a waltz by Tchaikovsky. I cannot hear that piece without being reminded of those tense seconds.

Only a selected group of Los Alamos citizens were allowed to witness the test. The rest were strictly forbidden to come near the Trinity site. To insure that only those with clearance would be there, it was announced that army patrols would comb the hills around the site. In spite of this, a number of people determined to witness the shot hid in the hills, waiting for the moment of detonation. When nothing happened at the planned hour, many of them went home. They thought that the whole enterprise was a failure. Many of those who stayed reported hearing suspicious movements during the night, which they thought were the army search patrols. But there were no patrols; the army had never gotten around to deploying them.

What the people heard were other would-be spectators trying to avoid detection. Those who stayed were rewarded with a spectacular sight just before dawn.

There were several official observation points. The nearest was a protected bunker six miles from the shot. The bunker contained most of the terminals of the observation apparatus and served as the test headquarters. Here Oppenheimer, General Groves, and other leaders of the project assembled.

Because I wanted to experience the full impact of the blast, I was at an open observation point ten miles away with Enrico Fermi, Phil Morrison, a few other physicists, and some of the military people. We were ordered to lie on our stomachs turned away from the center. Only the military personnel observed these rules. We scientists were not going to be cheated of our chance to witness the test in all its glory and frightfulness. Phil and I constructed a little frame of wooden sticks through which we planned to observe the size of the fireball in its initial phases in order to estimate the strength or efficiency of the blast, which was measured in equivalent tons of TNT. With our limited knowledge of the behavior of matter under such unusual conditions, we had predicted an efficiency of between 6,000 and 7,000 tons of TNT. The actual yield was almost 20,000 tons.

The countdown was heard over the radio—five, four, three, two, one, with Tchaikovsky in the background—and the sky lit up to an incredible intensity as the mountains flared under a light twenty times the intensity of the midday sun. Through dark glasses we saw a white hemisphere quickly expanding, with dust clouds around it on the ground. We were too excited to use our little construction for a good estimate of the size, as we had intended. After a few seconds we looked back at the mountains. They were still bathed in an intense white light.

The hemisphere grew, becoming a white ball that detached from the ground and slowly rose. All this took place in utter silence. Fifty seconds later a thunderous boom was heard reverberating several times from the surrounding hills. William Laurence, the *New York Times* reporter who was at our observation point, asked, "What was that?" because, having taken so long to reach us, the sound seemed unconnected to the explosion.

As the boom approached, Fermi stood there, erect and calm, slowly tearing little pieces of paper and letting them fall to the ground. He held a measuring stick in his hands. The displacement of air from the shock wave caused the slips of paper still falling to move horizontally by almost a foot. He measured the distance, looked into his notebook, and said, "Twenty thousand tons"—an example of Fermi's direct, simple approach to problems of physics. A more sophisticated instrument could not have measured the efficiency of the explosion more accurately.

The fireball grew larger and rose slowly, majestically into the air. As its light diminished in intensity, the protective glasses were no longer needed. Without them I watched the landscape fall back into the half-dark of early morning. We could now observe the changing colors with naked eyes. It was an overwhelming, awe-inspiring sight. A steadily growing ball, changing from white to yellow and then to orange, was rising slowly, with a stem of dust connecting it to the ground. The stem tilted slightly because a soft wind was blowing the ball to the right.

When the brightness subsided, we saw a blue halo surrounding the yellow and orange sphere, an aureole of bluish light around the ball. This effect of the radioactive radiation on the adjacent air was a totally unexpected phenomenon, although it would have been easy to predict. The appearance of this uncanny blue light made a deep impression on me. It reminded me, in spite of an inner resistance to such an analogy, of a painting by the medieval master Matthias Grünewald. Part of the altar piece at Colmar, the painting depicts Jesus in the middle of a bright yellow ascending sphere surrounded by a blue halo. The explosion of an atomic bomb and the resurrection of Christ—what a paradoxical and disturbing association!

The sky grew lighter. The fireball reached the layer of inversion, and it slowly transformed into a gray mushroom cloud. Finally the sun rose over the horizon, and its red rays illuminated the cloud and the oblique dust stem that still connected it to the earth. We had made it. There is no use denying the joy, pride, and satisfaction we all felt after that impressive testimony to our success. We had set free for the first time the immense cosmic forces hidden within the

atomic nucleus. It was ironic that our aim in all this was the development of the most destructive engine of death ever conceived.

We had estimated that thirty-six hours after the explosion the
ground radioactivity around the tower would have decayed enough
to permit a brief examination. Fermi, Bethe, and I drove in my jeep
to a point near the center. What we saw upon arrival was a flat area
about 400 meters in diameter in which the sand of the desert was
glazed into a solid reflecting surface. The tremendous heat to which
the soil was exposed had transformed it into a gigantic mirror by
melting the sand, which then solidified. The tower and all the huts
that we had constructed nearby were gone, literally vaporized into
gas by the heat. We learned later that when General Groves saw the
site, he exclaimed in some disappointment, "Is that all?" He probably
expected a deep crater leading to the center of the earth. General
Groves's disappointment could have been an argument against dropping the first bomb over an uninhabited region in Japan, since the
Japanese generals might not have been sufficiently impressed either
by the bomb's effects.

Our trip to the epicenter ended with a little scare. I had been given
an instrument called a "Victoreen meter" that measured the total
amount of radiation we received during the trip. In addition, each
of us carried a photographic film badge that also measured our radiation exposure. When we got out of the jeep at the edge of the
solidified molten sand, I looked at my meter. It read almost 200
roentgens, near the maximum of the scale. (A roentgen is a measure
of accumulated radiation. Six hundred roentgens are assuredly
deadly; 200 would represent a high risk of death. We had expected
no more than a roentgen or two from our exposure at the site.)

Terrified, I told my companions we had hit a "hot spot," and
urged them to run to the jeep. I did not reveal the threatening amount,
but I was deeply worried for our lives. At headquarters I hurried to
the medical center and showed the meter to the doctor in charge.
He had it examined by a technician, who declared it in working
order. In the meantime, our film badges were being developed, but
this process took time. In desperation I grabbed the meter and found
that there were two ways of putting the scale on the meter. Mine

had been put on incorrectly. The indicator was inserted backward, which caused a reading of close to zero to look like slightly less than 200.

After we were reassured that the radioactive cloud would not descend to inhabited regions around the site, we returned to Los Alamos. Our part of the job was finished. For those who were not involved in the delivery of the bomb over Japan, life in Los Alamos slowed down. We were all exhausted.

Oppenheimer had a friend named Katherine Page who owned a ranch in the Pecos mountains, and Ellen and I went there for a much-needed vacation. During the next few days we heard the radio reports of the destruction of Hiroshima and Nagasaki, followed by the Japanese surrender. We immediately became the center of attention. The other guests at the ranch considered us heroes who had defeated Japan. Finally we could speak openly about the project and tell the remarkable story of the release of the hidden energy of the nucleus.

In Los Alamos the mood had changed completely. The restrictions on discussions about the ethical and political problems raised by the bomb were no longer in effect. Anyway, it would have been impossible at this stage to shield us from all these questions. The secret of the bomb's existence—if not the technical details—was out, and almost immediately many of us were asked to describe our experiences in newspaper articles and in speeches. We talked about the tremendous power of the new weapon for evil and began to express the hope that this new power could help bring nations closer together. (Our hopes have not yet been fulfilled, but at least the ultimate catastrophe has so far been avoided.)

A unique period of our lives was over. We had spent only three years at Los Alamos, but they were three years of very hard work in the company of the best brains, the best people in my profession. We never again lived so intensely, nor did we ever learn so much. At Los Alamos we were enriched not only in our knowledge about our scientific fields but also in the human contacts we made. We lived intimately with a large group of people. All of us were engaged in a project that involved us intellectually, but many fellow scientists and their families also became cherished friends. And we were always conscious of our goal and of the dual nature of our achievement, which catapulted us into a new age in which the very survival of civilization was at stake.

Many of the deep friendships we established at Los Alamos continue today. Those three years irrevocably shaped our lives, our character, our worldview—and the world itself.

# NINE

⊠

# Return to Teaching, Research, and Politics

OUR GOAL HAD BEEN REACHED, but we entered a time of contradictions. Although we were suffering from the fatigue of grueling work, we had also grown used to the heady excitement and drama of our enterprise. We were proud of our achievement, yet we were burdened with the realization that we were responsible for creating the most destructive weapon ever devised. We lived with the awareness that our work had resulted in the deaths of several hundred thousand people under terrible circumstances, burned by the excessive heat and killed or maimed by radioactivity.

We had won the peace we had hoped for. But with the Allied victory and the end of the war came a series of paradoxes and moral conflicts it would take some of us many years to overcome. One question, in particular, remained unresolved: Why was the second bomb dropped over Nagasaki? The timing—just three days after Hiroshima—had not permitted the Japanese government to sue for peace had it wanted to. Could it have been that the U.S. military was eager to see the effect of a plutonium bomb now that they knew what a uranium-235 bomb could accomplish, even though our test at Alamogordo had shown that a plutonium bomb would work as planned? On some occasion I ventured to say that the first bomb might have been justifiable, but the second was a crime.

In any event, our task had been accomplished. Before we could be "discharged," however, the authorities asked us to prepare a report covering the past three years—not an easy assignment. The report was to be a systematic description of both our accomplishments and our failures. But since most of our decisions had been arrived at in conferences or discussions, and sometimes even only as the result of informal conversations, we faced a difficult task. Very few notes had been kept. So, in groups and individually we tried to reconstruct the details without benefit of documents that might have helped us recall the chronology of the process.

At the same time, we had become famous. After working so long in isolation and secrecy, we were suddenly in demand for talks, interviews, receptions, and celebrations of our work. The public regarded us as heroes for bringing the war with Japan to an end and saving millions of lives, and they clamored to hear us talk about our experiences at Los Alamos and to read the stories about our accomplishments. Perhaps we deserved the credit, but we can never know what would have happened had there been no bomb. Some historians have suggested that Japan was already seriously weakened to such an extent that the Japanese leaders might have asked for peace even without the nuclear attack, but that is conjecture. Certainly, in the immediate postwar period the Los Alamos alumni were credited with hastening the end of the conflict.

This resulted in a significant change in public attitudes toward science in general and physics in particular. Because physics and physicists had been involved in winning the war through the development of the bomb and the invention of radar, science was seen as having a greater impact on people's lives than nonscientists had realized. Suddenly the image of the obsessed—perhaps even slightly mad—man in an ivory tower engaged in mysterious experiments was relegated to literature and science fiction. Scientific achievements had become part of the general consciousness, and we were in demand to tell our story to a public whose imaginations had been fired up by our accomplishments.

We had a few months of relative leisure while the reports were being finished. Then it was time to return to our own laboratories and the research we had given up to build the ultimate weapon of war. I was not sure where I was going to go. MIT had engaged Professor Jerrold Zacharias to recruit new faculty for physics, with an emphasis on building a strong nuclear physics department. When

Zacharias offered me an associate professorship, including the possibility of advancement to full professor with tenure, I accepted immediately.

Ellen and I had always wanted to move to Cambridge. Its cultural attractions, including the famous Boston Symphony Orchestra, its museums, and the vast academic community promised a more interesting and active intellectual life than Rochester, although we were sad to leave our many friends there.

Living space in Cambridge was at a premium with so many people returning from the war, but we managed to find a house in nearby Arlington. Until it was available, Ellen and the children moved to Bass River on Cape Cod, while I lived in a furnished room in Cambridge and became a weekend commuter to the Cape. (We didn't move to Cambridge until 1953, when we found the house in which I still live today. It is one of the old Cambridge residences built more than a hundred years ago on an agreeable, quiet street lined with large trees.)

MIT offered some special programs that were unique at the time, such as the Research Laboratory for Electronics, the first of a projected group of interdepartmental laboratories. Norbert Wiener, a professor of mathematics at MIT whose work laid the foundations for today's communication and information revolution, shaped the laboratory's research program. Despite its name, the laboratory focused on almost anything related to human communication. A center for advanced electronics, it had expanded its scope to include studies of neuroscience, psychology, linguistics, and artificial intelligence. Eventually, the pioneering work of this institution spawned the cognitive sciences, establishing MIT as a leading institution in these fields and in research on how the brain works. It was here that Noam Chomsky and Morris Halle initiated their innovative approaches to the study of linguistics. All of this came out of a laboratory that during wartime had been devoted to the study of inanimate objects and whose work had resulted in the development of radar and similar weapons of war.

With the Laboratory for Electronics as a prototype, other interdepartmental laboratories were established, and I soon became active in the Laboratory for Nuclear Science (LNS). The range of this laboratory's interests included nuclear physics, nuclear engineering, and nuclear chemistry. Like the other interdisciplinary laboratories, it brought together researchers with related interests and resulted in a useful cross-fertilization of ideas. An enormous advantage of LNS was that its administrative staff was in charge of the financial support. We scientists helped frame the proposals, which included descriptions of our planned work, but the administrators wrote the requests for money, a job that would have robbed us of hours of precious research time. I never wrote a single grant request, a privilege few physicists can now enjoy. Much of the credit for this arrangement goes to Professor Zacharias, who was also the laboratory's first director.

Others who contributed to the special character of the labs were Julius Stratton, MIT's provost at that time; Walter Rosenblith, who later became provost; and Jerome Wiesner, president of MIT from 1971 to 1980. These men impressed me by the breadth of their vision. Rosenblith served as foreign secretary of the National Academy of Sciences for some time, and all three also worked for peace and international scientific collaboration. The existence of the laboratories helped MIT attract people with broad ranging interests and expertise. Moreover, the university had a reputation as a liberal institution that gave its staff great freedom. This proved helpful to me later. When I became CERN director in 1960, MIT gave me a generous leave of absence that covered the five years of my directorship. In other institutions I would have had to come back after two years or give up my professorship.

In one respect, however, I found MIT lagging. I was amazed that a major scientific institution could exist without a department of molecular biology. My admittedly amateurish interest in this area increased when the molecular structure of DNA, or deoxyribonucleic acid (the molecular basis of heredity), was discovered in 1953. Niels Bohr's involvement in these matters and my long friendship with the physicist and molecular biologist Max Delbrück were additional factors in my finding MIT's omission of this field to be a serious shortcoming.

It must have been around 1955 that I had a serious talk about MIT's biology department with Provost Stratton. I pointed out that molecular biology was developing rapidly but MIT had no part in this growth. I suggested that some experts in the field be invited so that they could advise the MIT leadership. Stratton took up my suggestion, and a number of young, active biologists were subsequently brought to MIT, including Cyrus Levinthal, Salvador Luria, and Alex Rich. Today MIT's biology department has an excellent reputation in this field, and I am proud of my small role in bringing this about.

Ellen and I soon made many good friends in our new community. One of these was Elting Morison. Elting, a historian like his uncle, Samuel Eliot Morison, came to MIT in the 1950s to teach literature and history to students who were basically interested in science. Before his first MIT lecture, Elting asked his students for a list of the ten books that had made the greatest impression on them. To his dismay, most of them had not read ten books. I don't know what would happen if the question were asked of today's students, but I'm afraid the response wouldn't be very different.

Elting was also a great friend of Stratton, who as provost was in charge of the general intellectual development of the institution. Stratton, who had a Ph.D. from the Swiss Institute of Technology in Zurich, said one day, "The spirit of Zwingli looks over this place." Huldrych Zwingli was an excessively puritanical Swiss Protestant reformer, and MIT was indeed a place of stringent intellectual endeavor directed at science and technology. To provide an antidote to this narrow atmosphere, at least for some of the faculty members, Elting founded what came to be known as the Friday Night Supper Club. A dozen people from MIT and Harvard were invited to gather on the first Friday of each month in Boston's Back Bay at the St. Botolph's Club, where Stratton was a member and the food and wine were excellent. The supper club was meant to provide a forum for stimulating, informal conversation in a comfortable atmosphere.

The members brought an eclectic mix of disciplines and viewpoints. Apart from Stratton and Morison, there was the Harvard economist Vassily Leontieff, then dean of Harvard College McGeorge Bundy, the historian Robert Wolff, the art historian Myron Gilmore, the sociologist George Homans, and the physicist Edward Purcell, all from Harvard, as well as the economist George Millikan and the engineer William Hawthorne, both from MIT, the Polaroid industrialist Edwin Land, and myself. Our unstructured discussions never followed an agenda. Sometimes I felt there was too much emphasis on internal university affairs, and the conversation too often deteriorated into mere gossip, but the discussions usually were wide-ranging. We talked about the political scene, philosophical topics, and occasionally the latest developments in our fields. For example, Purcell or I sometimes reported on exciting new results from physics research. Sometimes, particularly interesting visitors to Cambridge were asked to speak to the club, but on the whole, the topics developed informally out of our freewheeling conversations.

When I left MIT for CERN in 1960, I discovered that I missed the supper club, and I tried to arrange my visits to the United States to coincide with the meetings. I was always invited to report about my activities and the research at CERN and about the international scientific world. In the 1970s, however, the discussions increasingly turned to political issues. Indeed, politics eventually caused the demise of the club. During the Vietnam War crisis the members were sharply divided on the issues. McGeorge Bundy was an advisor in both the Kennedy and the Johnson administrations, and some of us differed sharply with his views on Vietnam. (Later, in the 1980s, Bundy became more liberal in his views about arms control and relations with the Soviet Union.)

In 1970, when Bundy was president of the Ford Foundation, he invited the other club members to New York to discuss our differences. They proved irreconcilable, however, and the club disbanded. Perhaps the times were no longer conducive to our kind of unfocused discussion group. Passions and opinions overpowered our discussions about heady subjects of intellectual interest. It seemed irrelevant to talk about abstract concepts while the country and our campuses were in such ferment.

During the seventies and even into the eighties Elting Morison organized occasional nostalgic reunions for the members and our wives, where we reminisced about the wonderful days of the old supper club, but these reunions could not compare with the actual meetings. Today there is a wonderful naturalistic statue of Samuel Eliot Morison on Commonwealth Avenue outside the St. Botolph's

Club, which he helped found. For me the statue is also a memorial to our much smaller organization.

Political and international issues affected my academic work, also. After my years at Los Alamos I had looked forward to returning to my old love, teaching, and to addressing the fundamental questions that still faced physics. But the consequences of my work on the atomic bomb and its impact on the world of the future weighed on my conscience. I found myself thinking often of Niels Bohr's ideas on this subject, in particular his great hope for an international management of nuclear matters as the only means of preventing an arms race that could otherwise lead to the ultimate catastrophe. Along with many of my colleagues, I gave speeches and wrote articles about these matters, and I helped found the Association of Los Alamos Scientists (which later became the Federation of American Scientists) and served on its council for several years.

The organization had several goals. We tried to educate the public about the importance of internationalizing atomic matters for the preservation of peace. We reiterated our belief that the monopoly on atomic weapons then enjoyed by the United States would be short-lived, publicizing our prediction that the Soviet Union would have nuclear bombs within a few years. Any industrial nation would be able to construct an atomic bomb; the basic process was well known. We were convinced that only by working together with the Soviet Union on nuclear matters could we save the world from a deadly nuclear conflict. We tried to make it clear to the public and to Congress that a nuclear war would be the ultimate catastrophe—that it would kill many millions of people and destroy the fabric of our civilization. A group of us also founded the *Bulletin of the Atomic Scientists,* which was then the only monthly periodical dealing exclusively with issues of nuclear policies, arms control, and disarmament. (Now it also addresses environmental problems.)

Another urgent concern was the administration of domestic nuclear matters, quite apart from the issue of international cooperation. Congress was considering two bills dealing with this subject. The May-Johnson bill would have put the management into military hands. Our organization favored the McMahon bill, which proposed a civilian Atomic Energy Commission. We lobbied congressmen and organized public groups in favor of this bill because we viewed civilian control as essential. The McMahon bill was passed, but, after forty years, I am not sure that passage of the other bill would have made as great a difference as we had feared. The Atomic Energy Commission supervised the construction of the nuclear bombs, but the decisions about the types and numbers and their deployment continued to be in the military's hands.

We had less success convincing the public about the danger of a nuclear arms race. Most Americans clung to the notion that the United States would continue its atomic monopoly, and in spite of the horror of Hiroshima and Nagasaki, they thought it was possible to defend the country against a nuclear attack. As an active member of the council of the Federation of Atomic Scientists, I made our views known in talks, articles, and interviews. But I was reluctant to participate in our contacts with congressmen. In retrospect, I am sure that one reason for this was my increasing immersion in physics; the other was my feeling that American politicians might be suspicious of my foreign accent.

Still, I followed the events avidly, and I continued to be particularly concerned about the efforts to internationalize atomic energy. In the beginning the negotiations looked promising. The United Nations established a committee, under the chairmanship of Hendrik A. Kramers, to draw up plans for an International Committee on Atomic Energy. Several such plans were submitted by U.S. spokesman David Lilienthal, who had the help of Oppenheimer and like-minded people. These ideas encouraged international management and included some serious restrictions on national decision making. Evidently, resistance from conservatives influenced the American delegation, and the plan was subsequently watered down to make it acceptable to Congress. In its final form it was known as the "Baruch Plan," named for Bernard Baruch, then America's elder statesman, who had been asked to draft a less radical version of an international agency.

Since the Soviets had not succeeded in constructing a nuclear bomb at that time, all of these plans included a transfer of pertinent information to the USSR. As expected, such intentions aroused the opposition of many people who did not want the West to give up its monopoly on nuclear weapons. The military, including General Groves, thought it would take the Soviets at least twenty years to catch up to American technology in nuclear weapons. It took them less than four.

In the end, of course, nothing came of our efforts to internationalize atomic matters, and we now can appreciate our naïveté. The mentality of the Stalin regime fundamentally opposed any collaboration with the West and any restriction on its sovereignty over military and industrial matters. The United States and Great Britain were not ready at that time to give up their temporary advantage in atomic development. All our idealistic hopes came to nothing. Our predictions were correct, however. The Soviet Union produced its first atomic bomb in 1949. As Niels Bohr had foreseen from the start, this event marked the beginning of a nuclear arms race. Not until the late 1980s did the superpowers, recognizing the madness of the situation, begin to initiate a few measures to reduce this international competition in arms.

Many of us recognized that the invention of the hydrogen bomb would make the situation much more dangerous. The H-bomb, using the energy resulting from the fusion of hydrogen nuclei with helium, would be many times more powerful than the old-fashioned A-bomb. I did not have to struggle with the problem of where to stand on the H-bomb development. For me such questions had long been resolved. After Los Alamos, I refused to have anything to do with nuclear weapons development. Never again would I be tempted to join in a project that would use my scientific knowledge to fashion a weapon of mass destruction.

Among those others who denounced the hydrogen bomb as being far beyond the limits of ordinary warfare were Oppenheimer, Fermi, Conant, and Rabi, who, as members of an advisory committee to the Atomic Energy Commission, proposed negotiating a treaty forgoing the development of this weapon in the United States and in the Soviet Union. Given the cold war climate of those times, it was not surprising that their attempts failed. Nevertheless, in 1949 a group of twelve physicists published a statement making clear our moral objections to such a bomb. In part, the declaration read:

We believe that no nation has the right to use such a bomb no matter how righteous its cause. This bomb is no longer a weapon of war, but a means of extermination of whole populations. Its use would be a betrayal of all standards of morality and of Christian civilization itself.

I remember vividly a conversation I had with Hans Bethe at his home that continued as we drove from Princeton to New York in October 1949. During the hours we spent together, I thought I had convinced Hans not to accept Edward Teller's request to help him find a way to make the H-bomb work. Unfortunately, two events in early 1951 influenced the future of the H-bomb: First, Teller and Stanislaw Ulam found a relatively simple way to make the weapon. This made it a virtual certainty that the Soviets would also soon be able to construct it. Second, the Korean War broke out. The North Korean invasion was seen as Communist aggression, and the fear of Communist world domination rose again. Under the influence of these two events, many physicists who had previously expressed doubts about the H-bomb began to participate in its development. Among them was Hans Bethe. Indeed, in 1951 both the United States and the Soviet Union succeeded in constructing this weapon almost simultaneously. It was successfully tested by both sides soon after, and the threat to humanity increased by another huge increment.

The Federation of Atomic Scientists, meanwhile, began to focus its attention on peaceful uses for atomic energy. We were convinced that nuclear reactors could become a cheap, virtually unlimited source of power, especially for the underdeveloped areas of the world. This vision inspired us to promote the harnessing of the atom for energy as a way toward a more peaceful and prosperous future. But time has proved us overly optimistic on that issue, too. We did not foresee that nuclear power would turn out to be almost as expensive as conventional power, and the dangers posed by malfunctioning nuclear reactors were not yet understood. We hardly even discussed the problem of how to dispose of atomic waste.

I still believe that nuclear power production will become necessary because the burning of fossil fuels pollutes the atmosphere with carbon dioxide, which may lead to a rise in the earth's temperature through the so-called greenhouse effect. Public opinion has turned against nuclear power in the wake of the accidents at Three Mile Island and Chernobyl. Nuclear power plants need to perfect sophisticated safeguards that will protect the population against nuclear accidents, but I am still convinced that safe nuclear power can be achieved and will be part of our lives in the not-so-distant future.

In the years just after the war, relations between the armed forces and the scientific community became friendly, since we had collaborated so well during the emergency. The scientists had proven extremely useful in the development of radar, the proximity fuze (an electronic device that detonates missiles), and the nuclear bomb. In response to this, the military began to fund scientific endeavors in peacetime. Unconditional support was readily available for practically any kind of research, including even the most esoteric investigations completely divorced from any possible military applications. Our war work had proven that physicists who spent their time on fundamental problems could successfully switch over to practical military tasks even without any previous experience in weapons research.

The Office of Naval Research was particularly generous in its grants, even if the work had no military application. It supported fundamental physics research at MIT, and we never felt any restrictions in our choice of research objectives. The situation has changed in recent decades. Today it would be more difficult to get financial support from the military for work with no direct weapons applications.

After Los Alamos we were all eager to start being pure scientists again, but as I have suggested, it was not altogether easy for us to shift from war work to fundamental physics. After an interruption of more than three years, many of us found it harder than we had anticipated to resume the work we had left unfinished. This was what inspired Oppenheimer in spring 1947 to organize a conference for physicists who had been working on the fundamental problems of the structure of matter. We were called together to identify the most important problems then facing physics. The meeting was to take place on Shelter Island, a lovely, isolated community on the eastern tip of Long Island, in New York. We assembled at La Guardia Airport and boarded a chartered bus. The mayor of Shelter Island had two sons who had been stationed with the army in the Pacific. Both had come home safely, and the mayor gave us—the developers of the atomic bomb—the credit for saving their lives. He sent his police force to escort us into town. We raced through Long Island to our destination, a charming little hotel set in the lovely countryside.

As it turned out, our return trip was also eventful. When the conference ended, Oppenheimer rented a private seaplane so he could go to Harvard to hear his good friend General George C. Marshall speak about his ideas for the economic revitalization of Europe. Oppie generously invited everyone who was going back to Cambridge to come with him. We took off over the ocean on a beautiful, clear day, but as we approached Boston, the weather grew threatening. The pilot, worried about flying into the storm, decided to land at the seaplane airport in New London, Connecticut. Unfortunately, this was a navy airport. The tower at New London informed our pilot in the strongest terms that civilian planes were forbidden to land there, but the pilot felt he had no other choice. As he came closer and closer to the landing site, we could see a fat, red-faced man waving and shouting at us, obviously signaling that we were not allowed to land. The pilot expressed his anxiety to us about the whole affair.

Oppenheimer patted him on the shoulder and said, "You just land the plane and let me handle this." When we touched down, the fat man, who turned out to be a captain, was furious with us. Sputtering and shouting, he told us we were breaking the law and described the penalties for what we had done. Oppenheimer came out of the plane first. "My name is Oppenheimer," he announced, and then he explained our reasons for trying to land. The captain gasped and asked, "Are you *the* Oppenheimer?" Oppie replied, "I am *an* Oppenheimer."

Then the captain realized with whom he was dealing. Instantly the mood changed from rage to veneration, and suddenly the captain couldn't have been more polite and cordial. He was terribly impressed that Oppenheimer, the great hero, had honored him by dropping in on his little field. Ceremoniously he led us to the command office, where we were given tea and cookies and put on a navy bus that took us to Boston. None of us was ever quite that famous again.

The conference at Shelter Island turned out to be exciting and important. Every day we gathered to talk about the problems in physics we had neglected during our years at Los Alamos. As a member of the planning committee, I organized the discussion on the difficulties of quantum electrodynamics, that is, the interaction of electrons with their own electromagnetic fields. My job was to recruit speakers, and I was asked to give a talk myself.

A sensational development on the opening day catapulted quantum electrodynamics into being the center of interest at the conference. At the first session, Willis Lamb announced that he and his collaborator, Robert C. Retherford, had reliably measured a small shift of a hydrogen level away from the values it should have had according to the fundamental equations of quantum mechanics. (This is the Lamb shift to which I referred in chapter 6.) The measurement had just been made possible by new technological advances in microwave techniques developed as a by-product of radar. The most plausible explanation for this shift was the interaction of the electron with its own electric field, but nobody had been able to calculate it at that time.

Naturally, the results rekindled my interest in these problems. Quantum electrodynamics was at that time a very successful theory. It correctly predicted the emission, absorption, and scattering of light by atoms; the existence of antimatter; and the production and annihilation of pairs of particles and antiparticles. But the interaction of the electron and the proton with the electromagnetic fields in their immediate neighborhood led to disturbing results. The energy of the electron (or its mass, according to Einstein) turned out to be extremely large when the effects of the fields very close to the electron were considered. The closer one got, the larger the mass appeared. This, of course, is in contradiction to the observed mass, which is rather small.

Furthermore, as I had realized in connection with my work in 1936, it also leads to one extremely large value of the charge if measured in the immediate neighborhood of the particle. At somewhat longer distances the large value is balanced by the opposite charge induced around the original charge. A close analysis of that result showed that the Lamb shift could be traced to the interaction of the electron with its field.

I felt guilty that I had not tried harder to calculate that shift before its final discovery. In principle, I could have done it in 1936 and certainly later by following some of the suggestions that Kramers had given me in 1940. He had argued that the difficulty of the extremely large mass (or energy) of the electron might be circumvented by comparing the energy of the free electron to that of the electron bound in the hydrogen atom. Perhaps the difference between these two very large quantities remains small because the large amount comes from the fields at very close distances from the electron, where most probably the theory is not applicable. These fields would be the same for the free and for the bound electrons. This difference would presumably be observable as a slight shift of some levels in the hydrogen atom. The difficulty lay in finding a reliable method for subtracting two very large quantities.

Shortly after arriving at MIT, I had started working on this calculation with Bruce French, a graduate student. Small shifts of spectral lines in hydrogen were suspected, but they had not been measured with any accuracy, and little was known about a shift of energy levels in hydrogen before Lamb and Retherford's announcement at Shelter Island. Unfortunately, in those postwar days at MIT, I had other interests, especially in nuclear physics, so I didn't vigorously pursue the investigation. This may have been a big mistake. Had French and I worked harder on solving this problem, we might have finished the calculation before Lamb announced the result of his measurements. What a triumph this would have been for theoretical physics. I might even have shared the Nobel Prize with Lamb, who received it for his work. Nevertheless, when French and I calculated the Lamb shift by using the laws of quantum electrodynamics and found that our results agreed with the experiment by Lamb and Retherford, we succeeded in showing that quantum electrodynamics is the fundamental description of electromagnetic processes, in spite of the fact that it still seems insufficient to describe what goes on in the immediate neighborhood of electric charges.

The discussion at Shelter Island initiated an important development in theoretical physics that led to a partial solution of the problems connected with the excessively large masses and charges appearing in the interaction of the electron or proton with its field. Shortly after the conference Hans Bethe published a brief paper in which he showed that the Lamb shift is indeed explained by the interaction of the electron with its field, although he could not yet get an exact result. Then French and I finished our calculation of the slight change in the interaction energy due to the binding of the electron in the hydrogen atom. Our figures agreed well with the measurements of the Lamb shift.

Possibly we were the first to make an accurate calculation. But I didn't want to publish the result because I had shown it to two distinguished theoretical physicists, Julian Schwinger, at Harvard, and Richard Feynman, at Cornell, who were very interested in our work. They tried to recalculate it and, working separately, both arrived at the same result. But their result differed from ours by a small amount. I was very upset by this development. If Feynman and Schwinger both got the same result and it was different from ours, we had most probably made a mistake. We began to recalculate and checked our results, but we couldn't find a mistake in our calculation. Half a year later Feynman called to tell me that he and Schwinger had both made the same mistake and that our result was the correct one, and he apologized publicly. In a subsequent paper he wrote in a footnote (appropriately numbered 13), "The author feels unhappily responsible for the very considerable delay in the publication of French and Weisskopf's result occasioned by this error."* In the meantime, Norman Kroll and Willis Lamb published a similar calculation with the correct result. Thus, we were not the first ones to publish, and they received some of the credit. All of this showed us the importance of having more self-confidence and of pursuing a project whether experimental data are available or not. In any event, it was an exciting experience to have done the work and to have arrived at the correct result.

The most important consequence of the Shelter Island conference

*Physical Review 76 (1949): 244.

and of subsequent ones organized by Oppenheimer in the next several years was the development of an elegant new method of dealing with the troubles arising from the fields at close distances from charges. It was much simpler and more general than the methods used by French and myself or by Kroll and Lamb. The method, called "renormalization," was worked out by Feynman, Schwinger, and, independently, by Sin-itiro Tomonaga in Japan. It allows one to calculate straightforwardly all kinds of interactions of an electron or of any other electrically charged particle with its own or another electromagnetic field. It was the final elegant realization of the vague ideas that I had published in 1936.

The idea is to separate the troublesome effects of the fields in the immediate neighborhood of the electron from all other effects, which are calculable with great accuracy. The Lamb shift is one of them. The theory was found to give an excellent account for many other effects, such as a small addition to the magnetic strength of the electron. The great success of the theory is based on the remarkable fact that only the mass and the charge of the electron are influenced by what is going on in the electron's immediate neighborhood, as I had postulated in my 1936 paper. That is why the theory can predict all the properties of the electron and all the processes in which it is involved, with the exception of the values of the mass and the charge. Those are, of course, known and well measured, but their values cannot yet be explained by the theory.

I did not contribute to this refinement of the theory, which gave it an elegant and simple mathematical form. Today nobody would use the clumsy method by which French and I—or Lamb and Kroll— calculated the Lamb shift. It illustrates my different approach to theoretical physics. I never was a good mathematician, and the formal elegance did not turn me on. I was satisfied when I understood the essential physical connections. This may have been a weakness in my way of looking at problems, since the formal beauty and neatness of a theory are very important for the further development of the fundamental ideas.

Another problem that led to lively discussions at Shelter Island involved the strange properties of mesons. These particles appear when cosmic rays penetrate the atmosphere or some other matter.

(Today they are produced in large quantities when highly energetic particles hit some material at the target of accelerators.) Mesons seemed to behave in an unusual way. Cosmic rays hitting the nuclei in the atoms of the atmosphere produced a large number of mesons. On the other hand, when these mesons traveled on through the atmosphere or through material, they seemed to be only very weakly absorbed by the nuclei they encountered.

This behavior was hard to understand. High production rates would imply strong interaction between mesons and nuclei, which should also have caused strong absorption. This widely discussed problem led to a paper by Fermi, Teller, and myself at the beginning of 1947 that pointed out how serious the discrepancy was. Several people tried to explain it, including me, but the correct explanation was given only by Robert Marshak at the Shelter Island conference and independently by Shoichi Sakata and Takesi Inoui, who had presented their explanations as early as 1943 before a Japanese symposium, unbeknownst to Western scientists.

They proposed that the meson produced in large quantities—now called a "pion"—is a different particle from the one that is weakly absorbed. The pion lives only a short time and then decays into another particle (now called a "muon"). Also a short-lived particle, it is very similar to an electron, but about 200 times heavier. In contrast to the pion, the muon interacts only weakly with nuclei, and this explains its weak absorption. Roughly at the same time, but unknown to Marshak and the others at Shelter Island, Giuseppe Occhialini, Cecil Powell, and their collaborators in Bristol were able to observe directly that the mesons created by cosmic rays do indeed decay into another particle, which turned out to be a heavy electron. This showed that Marshak's proposal was correct.

Another problem that engaged me at that time had to do with some work J. H. Van Vleck did at Harvard University on the broadening of spectral lines caused by the collisions between atoms. He told me about his calculations and discussed some of his problems with me; I was supposed to be an expert on line broadening because of my previous work on such problems. Perhaps I was of some help to Van Vleck, but I was astonished when he proposed to put my name on his paper as coauthor. I told him it would be enough to acknowledge my help at the end of the paper. He asked me, "Haven't you ever felt that your name should have been on a paper but wasn't?" I thought a little and said, "Yes, on a paper on the Lamb shift." "So," he replied, "accept your name on this paper."

I also published a paper on a similar subject with Aage Bohr, with whom I have had many pleasant and fruitful discussions on physics and politics. Somehow we have a similar approach to problems, and I profited greatly in discussing my own work with him and his close collaborator, Ben Mottelson, an American who settled in Copenhagen and became a Danish citizen.

One of the most attractive aspects of my return to the university was the resumption of teaching. I love teaching because it's the best way to learn a subject. Preparing a course reveals which points you think you understand but really don't. It has been my experience that, on the whole, the teacher learns more from the course than the student. Most of my research work suggested itself to me when, as I was teaching an advanced subject, I discovered that there was a great deal that was not yet well understood and in need of more thorough investigation.

I taught my advanced courses twice a week for one and a half hours, and I often used the first half hour to improve on what I had done in the previous class. Perhaps my teaching was in some ways too disorganized. I would try out one way of explaining something, and if I found it wasn't the right method, I would try again using another one. I like to teach in an informal way in order to show the students that things are often not as clearcut as they seem and that mistakes are often instructive. I spent a lot of time telling them, for example, what kind of difficulties one can get into, or what I first did wrong when I prepared the course. I think the students appreciated this; I'm sure the better ones did.

One of my most successful students, the Nobel laureate Murray Gell-Mann, made fun of my teaching methods at my retirement festival, saying that he and his fellow students found my courses very stimulating because they were always trying to catch me in a mistake. In order to find out what I really meant, they were often forced to go much deeper into the subject. I like to think that in this way they learned more than they would have in a better-organized course.

One of the advanced courses I gave after returning from Los Alamos was called "An Introduction to Nuclear Physics." I tried to put together what was known before Los Alamos with what we had learned there. It was wonderful to apply myself again to the fundamental questions irrespective of their relevance to technical problems. The preparation of the course revealed many holes in our understanding of nuclear reactions, and that led to numerous research problems. I joined forces with Herman Feshbach, at that time a young research physicist and assistant professor at MIT, and we set out to investigate a large number of problems that interested us both. We were particularly interested in understanding what happens when protons or neutrons collide with atomic nuclei.

Herman and I were able to work together productively because our approaches complemented each other. He had a profound mathematical understanding and much experience in solving mathematical problems. My thinking was more intuitive and directed toward qualitative pictures of what was going on. During the first decades of my stay at MIT, Herman and I, at first with two graduate students, Charles E. Porter and David Peaslee, published a number of papers that turned out to be important for the development of the theory of the atomic nucleus and nuclear reactions.

One of our approaches was called the "clouded crystal ball model." According to quantum mechanics, a particle beam exhibits properties akin to a wave motion. Therefore it is appropriate to use the analogy of a light beam (also a wave) to a particle beam. We described what happens when an energetic nuclear particle enters a nucleus by comparing it with a light beam entering a partially transparent and partially absorbing sphere. Indeed, a nucleus can sometimes be penetrated by an entering particle without any interaction, just like a light beam penetrating a transparent crystal. But it also happens that the particle is incorporated by the nucleus. It keeps sticking in it, as it were. The latter event corresponds to the absorption of a light beam in our model.

My interest in nuclear physics became so broad that in 1947 I decided to write a book on theoretical nuclear physics. As mentioned briefly in chapter 2, I selected as my collaborator John Blatt, one of the postdoctoral fellows at MIT. His way of working also complemented mine. Whereas I was more interested in the grand sweep of ideas and was somewhat casual about details and occasionally made careless mistakes, John was very careful and detail-oriented, and often caught errors that might have slipped by me. We laid out a plan for a treatise on theoretical nuclear physics covering the most important topics known at that time. We were able to finish it in five years. It became the most widely used textbook on that subject and is still used as a leading text in nuclear physics. I proudly report that for a time it was the book most often stolen from the MIT library.

Most people have no idea about how physicists work. If they give it any thought, they probably imagine the physicist as a person in a white lab coat, constructing or arranging complicated apparatus, using electronic measuring devices, peering at traces on photographic plates, or recording and comparing data. This accurately describes much of the day-to-day work of an experimental physicist, chemist, or biologist, but as I have mentioned earlier, a distinction between experimental and theoretical physicists has sharpened during the twentieth century.

The experimentalist must perform calculations of how the apparatus works and must also take into account how much accuracy can be expected. Of importance also is the ability to anticipate the causes of malfunctions that could simulate unintended results. A good experimentalist chooses experiments whose results will give better insight into the structure and dynamics of the object he or she is interested in.

The theoretical physicist, however, spends his or her time pondering the meaning of the results obtained by the experimenters. Most important, he or she must have a theory, an expectation of the results in a particular experiment, and must also be alert to what unexpected results would reveal about nature. The theorist spends all his or her time thinking about the object of investigation. Usually there are some preconceived ideas about the structure and dynamics of, say, an atomic nucleus. This idea may be expressible in a simplified model, like the clouded crystal ball model. Models are not considered an exact description of an object, but they can simulate its most important traits. If experiments show that the model correctly reproduces the traits, physicists gain a deeper understanding of the nature of the object. This is why Feshbach and I were gratified when Henry Barschall at the University of Wisconsin measured the effects of a beam of neutrons impinging on different nuclei and his conclusions agreed with those predicted from our clouded crystal ball model.

This description of the two types of physicist encompasses the typical representatives of the profession, not the few extraordinary geniuses. Ernest Rutherford was such a person in experimental physics. He chose experiments with an uncanny intuitive premonition that their results would bring unexpected insights. Several of them changed the course of physics.

In 1911 he exposed atoms to a beam of special particles (alpha particles) and discovered that the atom consists of a very small central nucleus surrounded by electrons. He had founded atomic physics. Eight years later he bombarded the nitrogen nucleus with a similar beam of alpha particles, and he transformed it into an oxygen nucleus. He had founded nuclear physics. A journalist once asked Rutherford why he always rode the crest of the wave of discoveries. He answered, "After all, I made the wave." Of course, the most famous geniuses of theoretical physics in the twentieth century were Niels Bohr, Albert Einstein, and P. A. M. Dirac. They went far beyond conceiving models or explaining some experimental facts. Einstein revolutionized our concepts of space, time, and gravity. Bohr created the most important concepts necessary for dealing with atomic reality. Dirac succeeded in unifying a relativity theory with quantum mechanics, leading both to the concept of antimatter and to quantum field theory, a consistent way of dealing with the interaction of matter with electromagnetic and other fields.

Today both experimental and theoretical physicists rely on computers in their work. Experimenters need them to collect and store the huge amount of data delivered by instruments and also to simulate the behavior of apparatus in order to assure themselves that these will do as expected. I have always been somewhat suspicious of theorists' use of computers. One can be fooled by computers. "Understanding" a theory means that one has a general idea of how the results follow from the theory. The computer is a help in getting more accurate and more reliable results. But I always advised my students to try to get a semiquantitative guess at the consequences of a theory before using the computer. This demands hard thinking but leads to a deeper insight into the theory.

The goals of both theoretical and experimental physics are twofold. The first is to discover phenomena and processes in nature previously unknown to us. The discoveries of X rays and of superconductivity are examples; the Lamb shift is a more recent discovery. This goal is reached mainly by experimentation, although sometimes new phenomena are predicted by theories, such as Einstein's prediction that a light beam is deviated by the sun if it passes nearby, and Dirac's prediction of antimatter.

The second goal is the discovery of the fundamental laws of physics, and for these discoveries theoretical physicists deserve most of the credit. Examples from the past include Newton's law of gravity and Maxwell's laws of electromagnetism. More recent examples include quantum field theory and the theory of the forces that hold the constituents of the proton (the so-called quarks) together, which is called quantum chromodynamics. My work with Feshbach is a lesser example. We wanted to get at the laws of the force that holds protons and neutrons together in the atomic nucleus and that gives rise to the behavior described by the clouded crystal ball model.

Whenever I join another community of physicists, I am struck by the close personal relationships that quickly develop among the majority of the members of the group. I cannot help feeling that the world of physics is made up of people who, taken together, constitute a kind of family. Over the years this has been demonstrated to me repeatedly as I have developed many lasting friendships with my colleagues.

My enduring friendship with Sid Drell began with a telephone call from Felix Bloch, who was then at Stanford, telling me about a gifted young man who had just received his Ph.D. Bloch asked me to take him on as a postdoctoral fellow at MIT. I had to say that all the positions in my department were already filled. But, Felix continued, this fellow is not only an excellent young physicist; he also plays the violin beautifully. I don't know whether Felix was aware that I had just lost the violinist with whom I played chamber music. I told Felix to give me a few days. In my talk with the department head I praised Drell, emphasizing his good recommendation from Bloch, but thinking about that violin. Somehow we managed to squeeze in another position, and as Felix had promised, I gained not only an excellent physicist but a very able musician. Sid contributed to the theory of elementary particles and was a stalwart fighter for arms control and nuclear peace. He and his wife, Harriet, quickly entered our circle of friends.

In 1957 I was able to convince the department to offer Francis Low a professorship, and we were lucky that he accepted the offer. Low contributed some fundamental ideas to particle physics, and his versatility in both physics and science in general was a great asset to me and the other members of our group. He was always eager to discuss any issue of physics or science policy. I loved to hear him say, "Viki, I have a problem . . ." as he was entering my office. Later he became MIT's provost. Both he and Sid Drell eventually succeeded me as chairman of the High Energy Advisory Panel to the U.S. government.

Some of my doctoral students also became good friends. David Frisch and his wife, Rose, worked in Los Alamos and came to MIT with me. He completed his Ph.D. under my supervision and then became a member of the department. Our devoted friendship, now forty-seven years old, is held together by our mutual liberal political views and by Dave's wonderfully humorous approach to life.

Certainly my most famous graduate student was Murray Gell-Mann, who received the Nobel Prize for his ideas about the substructure of the proton. At times I had difficulty keeping up with his quick mind. Nevertheless, I believe he did profit from our collaboration. I hope that I was able to transfer some of my attitude toward physics to him. Today he is a great authority on fundamental physics and is also deeply involved with finding solutions to the world's environmental problems.

Another student with whom I had strong personal ties was Francis Friedman. After working with me on some problems of nuclear physics, he devoted most of his time to pedagogical concerns and was the leading voice in a broad attempt, organized by Jerrold Zacharias, to reform physics education in high schools. I always approved of efforts to improve science teaching on any level. Unfortunately, I could not actively participate in this particular project because I left for CERN at that time.

Finally, David Jackson and Kurt Gottfried belong in the group of my students with whom I remain in close personal contact. I think I was able to share with both of them my interest in the clear and simple presentation of modern physics. Of course, such influence is possible only if the recipients have an innate bias in that direction. Apart from their successful research, both David and Kurt wrote well-known textbooks, David on electricity and magnetism and Kurt on quantum mechanics. With Kurt, I shared my concerns about the nuclear arms race, and he became more active and knowledgeable in that area than I. We also shared our love of the mountains, but both Kurt and Dave outclassed me in their skills and were able to embark on many climbing expeditions for which I lacked the training.

These friendships were forged in Cambridge. Many more were made when I went to Europe later in my career.

⊠

# TEN

# Growing Ties to Europe

ELLEN AND I WENT to Europe nearly every year from 1947 to 1960, usually during the summer, and we spent my two sabbatical years there. Although we were by then thoroughly integrated into American life, we remained conscious of our European origins. I was in Zurich for a summer semester in 1948 and in 1950. When I got a Fulbright scholarship for the 1950–51 academic year, we spent the fall term in Paris, where I was appointed Visiting Professor at the Sorbonne. We moved to Zurich for the spring term.

When we arrived in Paris at the end of September 1950, we were surprised to find that the semester didn't start until November. We brought our children to the Chalet Flora, a French-language boarding school in Gstaad, and took advantage of the free time to travel in Switzerland and Italy. We didn't tell anyone where we were going. In October we read in the papers about the disappearance of the Italian physicist Bruno Pontecorvo, who had an influential position in England. He soon turned up as a defector to the Soviet Union. When we read this news, we thought that our disappearance from Paris might also be interpreted as a flight to the East, and we hastened to send a telegram to Paris assuring my colleagues at the Sorbonne that we would be back for the start of the lectures.

From Tom in Gstaad we received some disturbing mail. He was ten years old and had always liked to learn new things, but his letters complained that when his teachers spoke, he just heard a lot of noise—he couldn't understand a word. We were very worried that he was going to be desperately unhappy there. But after only about three weeks the "noises" had been transformed into recognizable words, and his letters became cheerful. By the time we saw them again, both children were on their way to speaking French without an accent.

Unlike my children, I experienced some problems with the language. In Paris, I tried to lecture in French, which was not too easy for me. I had learned the language as a child in Vienna from a French governess, so I knew "child French" very well—but that is not exactly the language in which to teach nuclear physics. For example, when I had to speak of free neutrons, those moving freely in space without any forces acting on them, the French word for "free" didn't come to mind, so I said *neutrons frites,* which means "fried neutrons." This became a constant joke among the students. But on the whole it was a good exercise. Every day my French improved, and this helped me later when I became director of CERN in Geneva, where French is the spoken language.

Just before beginning to teach in Paris, I had finished my book on theoretical nuclear physics, so I tried out the text in my lectures to the students. The lectures were very well received. Nuclear physics was just beginning to be taught, and everyone was eager to hear about it. As I mastered French, I quickly developed very good relationships with the students and the staff.

The Chalet Flora was wonderful for the children; they had a good life. We visited them for the holidays and for skiing vacations in the winter, and they loved staying with us in our hotel. Once we went out for dinner with them, and our daughter refused to eat the peas that were served. My wife reminded her that at Chalet Flora she had to eat them. Our eight-year-old said, "If you *have* to eat them they don't taste so bad." I always thought that was a good philosophy.

Our time in Zurich in 1948 and 1951 evoked pleasant memories of Zurich in the thirties. The difficulties with the *Fremdenpolizei* were forgotten, and we enjoyed the sedate life of that city with its rich cultural resources. Of course, the theater was not as unique as it had been when it was staffed with the best German refugee actors, but life was comfortable and restful. We renewed our relations with friends from the old days and made new ones, especially among the foreign students at the university and the famous ETH (Eidgenös-sische Technische Hochschule, or the Federal Institute of Technology). In the first postwar years many students came from abroad to Switzerland, which had been spared the ravages of war.

I always like to be in touch with graduate students; you find out more about the physics that is going on and also about the foibles of the professors. Three young postgraduate students impressed me particularly. Valentin Telegdi, from Hungary, and Amos de-Shalit, from Israel, were experimental physicists working toward a Ph.D. in the laboratory of Paul Scherrer; Igal Talmi, a theorist, was working with Wolfgang Pauli.

Students usually get instructions from the professor for the experiments they are supposed to carry out, but both Telegdi and de-Shalit suggested ideas not only for their own work but also for that of other students. Later I invited de-Shalit to come to MIT to work with me on theoretical problems. I wanted to get Telegdi too, but there were not enough open positions. I recommended him to the University of Chicago, where he did excellent work and in due course became Enrico Fermi Distinguished Professor of Physics.

De-Shalit and I developed some new ideas of nuclear matter, that is, the substance consisting of protons and neutrons, of which the nuclei are formed. He returned to Israel after his stay with me and became one of the intellectual and administrative leaders of Israeli physics. He served as director of the Weizmann Institute of Science in Rehovot until he died prematurely in 1964.

Val Telegdi is unique among experimental physicists. His understanding extends beyond the theoretical basis of his experiments to the intellectual content and ingenuity of experimental methods. It is the rare teacher who can convey to his students an idea of the kind of elegance essential to carrying out experiments. Val did just that in a series of lectures at MIT entitled "Great Experiments." He would choose a number of recent decisive experiments and not only explain the theoretical background but in great detail describe the ingenious ways by which the best and most reliable results were achieved. I have never heard any other lectures or seen any book that gave such an impressive testimony to the intellectual values of experimental physics. I hope that some day he will find the time to write up these lectures.

Val is very demanding of himself and of others. If he condemns someone, the poor person may or may not be so bad. But anyone he praises is certainly outstanding. He despises loose statements about ideas that are only half understood. These strong opinions have gained him a reputation as a harsh and aggressive discussant. On the other hand, Val is one of the most loyal, considerate, and helpful of friends to those he values highly. Such persons are even permitted to indulge in the kind of loose and inconsistent thinking he usually condemns in others.

Telegdi's talents range throughout physics and far beyond. He possesses a phenomenal knowledge of the physics literature and has an unfailing memory. He knows the history of most relevant ideas, and he can quote the contributions of the most unlikely persons in the most obscure journals. Connected to his unfailing memory is his knowledge of languages. He speaks many difficult languages (such as Hungarian and the form of German spoken in parts of Switzerland) practically without an accent. Strangely enough, in English, the language he uses most frequently, he has a distinct accent. Because of his good ear for languages, he can imitate the idiosyncrasies of the way people speak with uncanny accuracy. Once he gave me a record of a discussion between Pauli, Scherrer, and Gregor Wentzel, the University of Zurich theorist. No one I played it for realized that all the voices were, in fact, Telegdi himself imitating these three distinctive and well-known accents. This was also an example of Telegdi's sense of humor. He has devised countless jokes and puns that have enlivened many boring scientific conferences.

I have always enjoyed spending time with Telegdi and his charming wife. It remains a rewarding experience for me after forty years of acquaintanceship to discuss my physics problems with him. His critical sense and his unfailing memory of the literature have often helped me in preparing publications or talks. I can't count the number of times he has pointed out inaccuracies, non sequiturs, and erroneous or missing quotations of other people's work. I sometimes refer to him as a living encyclopedia of physics. In addition, his love of good food makes him a superb guide to the best restaurants all over the world.

During those European summers we returned to Altaussee with the children. Our family's house had been confiscated by the Nazis after the Anschluss and given to a Nazi party member. He had subsequently sold it to a citizen of Altaussee, who was still living in the house. According to the postwar laws of reparation to victims of the Nazis, I had the right to take the house away from this man because it was property that had been illegally seized and sold. I couldn't bring myself to do this. I planned to spend many summers in Altaussee, and I was unwilling to do anything that would alienate the local residents. Instead, I resold him the house for a very reasonable price. We spent our Altaussee vacations in rented apartments.

To be honest, the house was no longer as desirable as it had been. It had been built rather close to the main highway; in those days, one traveled by horse-drawn carriage. Automobile traffic had changed the character of the place. We were comfortable in our rented apartments and had few regrets about the loss of my family's summer house. My greatest joy was watching our children as they discovered our mountains, meadows, and forests. They climbed the same rocks in the lake that we had climbed, and they loved them as much as we always had.

During the summers we were never completely on vacation. There were always invitations to conferences or lecture engagements. Gradually, our European sojourns became an integral part of our lives. We began to feel like "Atlantic citizens," at home on both sides of the ocean, and the children profited from being in touch with both cultures. Being back in Europe was important to us because we were eager to reestablish our deep cultural roots. Although our American life was a great asset, we had learned the value of the combination of the two traditions. In a play we had seen during our first years in America, a British immigrant says something like, "I know England like the palm of my right hand and I know America like the palm of my left hand, and I put the two together and pray." Somehow this seemed symbolic of our life. We had left Europe before the war, convinced that European culture was about to be destroyed by

Hitler. Fortunately this didn't happen. Now we were able to partake of both worlds.

We were happily surprised at Europe's rapid recovery after the war. Whenever we visited places of great culture, such as Italy, France, Germany, Holland, England, and the Scandinavian countries, we were profoundly moved by the enduring strength of the European spirit. And after a decade of interruption we again saw the familiar cathedrals when we visited the rebuilt cities, and went to the German-language theater performances like those that had meant so much to me in my youth. We came to realize that there is a Western culture common to America and Europe. This tremendous spiritual edifice was built on ancient Greek and Roman foundations and spread all over Europe, where it produced intellectual and artistic creations of enduring value. The great painters, composers, architects, and writers provided us with those things that make life worth living in the face of all-too-frequent oppression, cruelty, and conflict. I was keenly aware that the same creative ferment gave rise to the scientific insights of Copernicus, Galileo, Newton, Maxwell, Darwin, Einstein, and Bohr.

We often visited Ellen's homeland, where our children had no trouble becoming attached to the Danish scene. When I asked our son how he could communicate with the Danish children with whom he played, he said, "We can't talk together, but we can laugh together."

In Denmark our vacation time was enriched by the annual conferences on theoretical physics at the Bohr Institute. These had been started almost immediately after the war and were continued until Bohr's death in 1962.

There was much to discuss, including the new developments in quantum electrodynamics that had been reported on at the Shelter Island conference in 1946. Many new discoveries had been made regarding the properties of the atomic nucleus, such as the so-called shell structure of nuclei. This is one of the parallel features of the nucleus and the atom—the former made of protons and neutrons, and the latter of electrons and nuclei. Both systems, although different in size by a factor of ten thousand, are governed by the same laws of quantum mechanics.

185

At this time particle physics was still a new science, and one of the first surprising results came from early experiments with cosmic rays. New short-lived particles were found. Much experimentation and calculation were necessary before the nature of these new particles was recognized. Some were identified as excited states of the proton or the neutron. Some were found to be mesons; others were identified as the previously mentioned short-lived heavy electrons called muons. To everyone's surprise, these muons have all the properties of an electron but are 200 times heavier and decay into a normal electron and neutrinos in a few microseconds. The existence of heavy electrons—a third, the tauon, which is 3500 times heavier than the ordinary electron, was discovered in 1975—remains a great riddle even now. Why did nature come up with these new varieties of electrons? When the discovery of the muon was first announced at a meeting, Isidor Rabi exclaimed, "Who ordered that?" Today we still don't know.

Nuclear and particle physics were very lively and exciting sciences at that time. But discoveries and ideas in other fields of physics were also discussed at the Copenhagen meetings. For example, superconductivity, the ability of some metals to conduct electric currents without any resistance at very low temperatures, was for decades an unexplained phenomenon. An explanation was finally found in 1951 by John Bardeen, Leon Cooper, and Robert Schrieffer. Many other insights into the behavior of light and materials were among the startling developments and discoveries we discussed at that time.

Naturally, we revived the tradition of a comic session at the institute. Those of us who wrote the skits and put together the comic evenings before the war, to the great amusement of the older generation, had looked forward to the day when we would sit in the first row and see a show put together by a younger generation poking fun at us. But it did not turn out that way. Once again, George Gamow, Max Delbrück, Hendrik Casimir, Carl Friedrich von Weizsacker, and I had to think up the ideas and produce the skits. We had a wonderful time taking off on our own generation. Among our favorite targets were the scientists who quickly invented theories to explain observations that were still doubtful. For example, we showed a completely blank picture supposedly taken by a bubble chamber, an instrument that shows colliding and decaying particles. Casimir imitated the voice of a well-known theorist announcing importantly, "This is an event that has not yet taken place. But it is *not* too early to draw the following conclusions . . ."

Our comic performances were now given in English, which had replaced German as the scientific language of Europe. As spoken by Niels Bohr, however, every language sounded roughly alike. We called it "Bohrish." His accent was so impenetrable that often one couldn't tell what language he was speaking. Once, before the war, he gave a lecture before a large audience at a medical conference. The subject, light and life, dealt with his philosophy of the complementarity between life phenomena and physical and chemical phenomena. Delbrück and I were in the audience, and it took us ten minutes to decide what language he was speaking. At one of the Atoms for Peace conferences in Geneva, earphones were used for simultaneous translations. Bohr spoke in what he thought was English, but earphones were necessary in his case so that the translator could transform his words into English that could be understood.

During our frequent visits to Denmark, my friendship with Bohr grew deeper. Many of our conversations concerned world politics, the increasing intensity of the cold war, and the breakdown of his efforts to establish an International Nuclear Authority. We were constantly seeking ways for scientists to help improve the situation. When I was in Copenhagen, we met either in Bohr's office at the institute or in his mansion near the Carlsberg beer factory. We sat together in the very room in which I had been a "victim" in 1933. Our most pleasant encounters, however, took place in Tisvilde, a resort on the northern shore of Sjælland, a two-hour drive from Copenhagen. The Bohrs spent the summers there in their old Danish country house, which had a thatched roof and antique furniture. On the adjacent land they had built houses for the children and their families.

A hundred yards from the main house, Bohr had furnished a ramshackle barn with a few uncomfortable chairs, a table, and a blackboard. Here was where he loved to work and talk to visitors. I spent many wonderful hours in that barn, studying problems with Bohr. We would remain deeply involved in our conversations until

Margrethe called us to lunch. Then a procession made up of Bohr, the families of his children, and I streamed toward the old house to have a meal of the wonderful Danish smøorrebrød, coffee, and cake. It was a special pleasure for me to participate in the life of the Bohr clan in Tisvilde.

Due to my ties in European science, I became increasingly interested in scientific contacts and collaboration between the United States and Europe. Such contacts had been resumed immediately after the war, and since Europe afforded limited opportunities for frontline science in the aftermath of the war, many young European scientists were eager to work at American scientific institutions. At that time, anyone who wanted to have a successful scientific career in Europe almost had to have studied in the United States. Luckily, American institutions were remarkably generous in offering graduate and postgraduate fellowships to European applicants. After the war, up to 40 percent of the graduate students at American universities were Europeans. These young people established an excellent base for scientific collaboration when they returned to their homelands. Those who came with a European fellowship encountered some financial difficulties because of the high value of the dollar in the early days of the postwar era. Their stipends were usually augmented by American grants. (When my grandson recently received a Fulbright Fellowship, his European sponsors had to offer him additional income so he could live on his American grant. Times have changed.)

In the early 1950s the cold war cast a shadow on these encouraging developments, and the wave of extreme anticommunism promoted by Senator Joseph McCarthy cast a further pall over international scientific endeavors. I don't know why I never fell under suspicion during the McCarthy years. It was common knowledge that I had spent some time in Russia and had visited the Soviet Union frequently and that I had been a member of the Social Democratic party in Vienna. Perhaps I escaped censure because I had always been publicly critical of the Soviet system under Stalin.

A number of my colleagues at American universities had expressed Communist sympathies or had had connections with Communist sympathizers before the Second World War. Although most of them had changed their opinions by the 1950s, many came to the attention of the House Un-American Activities Committee and had to suffer painful interrogations. Those of us who had witnessed the Nazi oppression in Europe were disappointed to see how some of our colleagues revealed the names of their leftist associates when they were called before the committee just because they were afraid of losing their positions. We remembered those people who had resisted telling the Nazis of their associates under the threat of arrest and death in the concentration camps.

Of course, there were also some who behaved wonderfully and courageously, steadfastly refusing to give names. One of these was Wendell Furry, then a professor of physics at Harvard. It was well known that he had been rather closely connected with Communist circles before the war. When called before the committee, he withstood an unpleasant grilling without giving any names. Furry did not lose his tenured position at Harvard, in spite of great pressure on the administration to dismiss him. Unfortunately, some younger untenured professors who were attacked by McCarthy were not so fortunate and had to leave their institutions.

A more serious case involved Dirk Struik, a professor of mathematics at MIT. Struik also had sympathies with the Communist cause and with Marxist philosophy, although he was never politically active and never tried to influence his students. Struik was very strongly attacked by the committee, and a Massachusetts court indicted him for wanting to overthrow the constitution of the commonwealth of Massachusetts. Due to the hysterical climate of the times, every professor who wanted to teach at a Massachusetts university had to sign a ludicrous oath swearing that he or she would never attempt to overthrow the Massachusetts constitution. When the indictment was handed down in fall 1951, MIT suspended Struik but paid him his salary.

I joined a group of faculty members who objected to his suspension on the grounds that, as long as he was only indicted, he should be considered innocent, according to American law, and we asked MIT to rescind his suspension and grant him a leave of absence. We were careful to express our opinion in a way that was loyal to MIT while reflecting our sense of the injustice of this action. Our efforts did not prevent Struik's suspension, but we did succeed in one respect. The first announcement of his suspension excluded him from the grounds of MIT, including the library. Our group convinced the administration that this was going too far. Two months after the first announcement, he was given the right to confer with his colleagues and to use the library. Struik was never brought to trial. When the indictment against him was dropped five years later, MIT restored him to full status. He returned to teaching and to his regular activities as a member of the MIT community.

My friend David Hawkins was similarly hounded by the committee and put through a very disagreeable round of questioning. He held out and refused to name any names. Fortunately, he did not lose his job as professor at the University of Colorado.

The matter of Oppenheimer and the committee has often been described and discussed. Most of the members of the scientific community were very distressed when he was denied security clearance and brought to trial. Oppenheimer had been regarded with suspicion by rightist circles ever since he had proposed an agreement with the Soviet Union barring the hydrogen bomb's development by either that nation or the United States. Some of my colleagues testified for Oppenheimer when he was called before the committee. I was not called to testify, as I was not so well known. I did write him several letters expressing sympathy and support. In one of these letters I said,

> Somehow Fate has chosen you as the one who has to bear the heaviest load in this struggle. I know that you are suffering from this as any man would under such enormous strain. On the other hand, I would not know of any better man to bear this load. As a matter of fact, if I had to choose whom to select for the person who has to take this on, I could not but choose you. Who else in this country could represent better than you the spirit and the philosophy of our way of life? Please think of us when you are feeling low. Think of all your friends who are going to remain your friends and who rely on you.

There was one occasion in 1949 when Oppenheimer disappointed me. A Rochester, New York, newspaper reported statements made by him concerning Bernard Peters, a former student of his, then a professor at Rochester University, who was under attack by McCarthy. Reportedly, Oppenheimer told an investigating committee that Peters was a member of the Communist party and a proponent of violent action. Peters denied these allegations, although he admitted his sympathies for leftist movements. Hans Bethe, Edward Condon, and I wrote strong letters to Oppenheimer urging him to rectify his alleged statements. He eventually did, and although he defended Peters's right to have unorthodox opinions, Peters ultimately lost his job at the university. He left the country and took a position in India and later one in Copenhagen.

Particularly damaging to international cultural relations was the passage in 1950 of the McCarran-Walter Immigration Act. Although the act contained some important improvements regarding the distribution of quotas for immigrants, which previously had favored Anglo-Saxons, it severely curtailed travel of foreigners to the United States. Under the influence of McCarthyism, the act was restrictively interpreted. In order to obtain a visa to enter the United States, even for a short time, applicants had to fill out forms indicating their place of residence going back to the age of sixteen. In addition, the form required a list of all of one's past and current memberships in organizations, professional groups, societies, clubs, and political parties—including the date on which one had joined. It was also necessary to describe each organization's character and aims and the "function of the applicant" in each group. Obviously, these tasks proved virtually impossible. Any mere hint of leftist connections, such as having been seen with people of leftist orientation, could provide a reason for denying a visa. These decisions were mostly in the hands of the consuls, who feared making mistakes and therefore were always more ready to deny a visa than to grant it if even the slightest suspicion existed.

Well-known and distinguished writers, musicians, and scientists were often excluded from the United States for the flimsiest reasons. In April 1954, P. A. M. Dirac, by then one of the world's most famous physicists, was considered ineligible for entry from England. The scientific community protested forcefully, but it took until August for Dirac to finally win a visa, and by then he had to postpone his visit by a year because of other commitments. Several international scientific institutions decided not to hold meetings in the United States or in any country that had political restrictions on those who wanted to visit. The people who suffered most were the younger scientists in America. They did not have the means to go to foreign meetings and lost contact with the international scientific community.

I was, of course, particularly interested in using whatever influence I had to alleviate the situation. I considered this my duty to international science. At my suggestion the Federation of American Scientists established a committee for visa questions, and I became its chairman. Overcoming my earlier reluctance, I testified before Congress and emphasized the negative aspects of these measures on the reputation of the United States and on scientific progress. "If this law had been in effect during the thirties," I said, "many of the Hitler refugees who contributed so much during the war would not have been admitted. We would have been slowed down in the development of the atomic bomb, radar, and many of the other scientific breakthroughs that won the war for the Allies."

I was not the only one who spoke out, and eventually our actions had a certain influence. By 1955 the situation had improved a bit, and more people could come in. We were particularly successful in one important respect. Before the mid-1950s a foreign visitor invited to a U.S. conference had had to apply in person for a visa. If the visa was refused because of alleged leftist connections, the applicant had to file a request for an exception. The renewed application also had to contain an admission that the applicant was a leftist. Understandably, many foreign scientists were too proud to ask for an exception because of accusations about having had a conversation with a supposed leftist or because of some other innocent act that displeased the U.S. government. They preferred not to come at all. If I had been a foreign scientist, I probably would have had the same attitude. We convinced the State Department to accept requests for exceptions from the American institution that had invited the foreigner, instead of from the invited scientist. In many cases we alerted the State Department staff ahead of time as to which foreign visitor we wanted and asked them to inform the consuls that a visa was desirable in spite of some difficulties. This approach largely succeeded, and getting people here became easier.

While Stalin was still alive it was impossible to exchange visits between East and West, so Eastern and Western scientists were robbed of the opportunity for professional and personal contact. After Stalin's death in 1953, we attempted to establish some connections to our counterparts in the USSR. Two years later, at the invitation of the Soviet government, a group of twelve American physicists, including myself, visited the Joint Laboratory for Particle Physics in Dubna, the Eastern European equivalent of CERN, about a hundred miles outside of Moscow.

We arrived at the airport in Moscow and were met by a group of perhaps a dozen Soviet physicists, most of whom I knew from my previous visits to the USSR. Among them was my old friend Lev Landau. We embraced warmly after having had no contact of any kind for more than two decades. It was an emotional moment for us both. We went on to Dubna for the conference, and there, as we spoke to our Soviet colleagues, we were not surprised to find that we shared many interests and had worked separately in our own laboratories on many similar scientific problems without any personal communication. We were impressed by the state of Soviet science. It was more highly developed than we had expected, although it certainly was not on the level of Western science. Referring ironically to the boasts of the Soviets about the state of their science, which many Western reporters called lies, one of my colleagues said, "All their lies are true." This was an exaggeration, but it was remarkable how much the Soviets had been able to do with their limited means. And in theoretical physics, where the basic tools are pencil and paper, they were on a par with Western scientists.

The high standards of Soviet theoretical physics were then and are now due to a large extent to Lev Landau and his school. Landau must be considered among the best theoretical physicists of this century. His interest in almost all fields of physics was remarkable. He contributed seminally to many branches of physics, from quantum mechanics and field theory to condensed matter physics, and in particular to the theory of phase transitions and the superfluidity of helium. Landau had a deep physical understanding above and beyond mathematical formalism. It was always a pleasure to discuss problems with him, and I always ended up with better insights into the real significance of what we were discussing.

After the conference Landau invited me to his house in Moscow, where he had assembled a few of our old friends and collaborators from the thirties. When I visited Landau's bathroom, I found a volume of Stalin's autobiography where the toilet paper should have been. This was typical of Landau, who was famous for taking political risks. Although Nikita Khrushchev had just made his first public statement about Stalin's misdeeds, using Stalin's book for toilet paper would probably not have been considered approved behavior.

Unfortunately, an automobile accident in 1962 ended Landau's intellectual life. His brain was damaged in such a way that he could no longer read a book or a paper, much less formulate any ideas. He remained in a hospital until his death in 1968. I visited him there, and it was a harrowing experience. His short-term memory was destroyed, but he remembered everything about the past. We talked about our old days together; he not only recalled many details, he also spoke fluent German. But every three minutes or so he forgot that I had been with him for some time and said, "Oh, how nice of you to visit me here in the hospital." Landau's death was a great loss for Soviet physics and for the world.

In 1956 I and several others wanted to get permission to invite Soviet physicists to the United States to attend the annual Rochester conference, which my friend and collaborator Robert Marshak, then a professor at the University of Rochester, had initiated in 1950. It was considered a sequel to the Shelter Island meetings. Groups of physicists gathered in Rochester to discuss the latest developments on the frontiers of particle physics. About fifty or sixty physicists attended the first meetings, but by the 1970s and 1980s many hundreds of people came. Although they continued to be called Rochester conferences, the meetings were eventually held on a rotational basis in the United States, Europe, and the Soviet Union, and later in Japan.

The first two conferences, in 1950 and 1952, were attended mainly by American physicists; by 1953 the conference had become more international. Scientists from Great Britain, France, Italy, Australia, Japan, and several other countries were invited to this third annual conference. Unfortunately, because of the difficulties of the McCarran-Walter Act, several invitees could not come. When contacts with Soviet scientists began to develop in 1955, it was natural to think of inviting them, too.

The 1956 conference was the first opportunity for getting Russian scientists to come—a difficult task, given all the problems we had had even with scientists from non-Communist countries. My previous experience with visa problems made me the obvious choice as the person to try arranging matters for the Soviets. I enlisted the aid of the MIT president, James Killian, the former science advisor to President Dwight Eisenhower, thinking he might be able to give me both scientific and political advice. He suggested that I talk directly to the Central Intelligence Agency (CIA) staff, in particular to Dr. Richard M. Bissell, a highly placed officer of the agency, which was then under the direction of the very conservative Allen Dulles.

My colleagues and I tried to convince these people that it was vital—from many points of view, scientific and otherwise—to allow the Soviets to come to this conference. We wanted to reestablish a true international community of scientists to create a better understanding between East and West, which was at a low point at that time. Because some Soviet physicists were first-rate, especially in theoretical physics, we would learn much from them through personal contact. I encountered opposition from many of those in authority, in particular from Lewis Strauss, the chairman of the Atomic Energy Commission. I thought that the best way to convince him would be to ask Edward Teller and Eugene Wigner for help. These two gentlemen were politically conservative as was Strauss, but they shared our conviction that the participation of Soviet scientists in the Rochester conference would be desirable. We were finally able to convince Strauss after many letters and telephone calls and dozens of trips to Washington.

When we were at last granted permission to invite the Russians and were assured there would be no visa problems, we asked six of their best scientists, among them Lev Landau, to come to Rochester.

I was not surprised that the audacious Landau was among those who were not allowed to accept our invitation. In the end only Vladimir I. Veksler, Moishe A. Markov, and Victor P. Silin came, but it was a great triumph. For the first time since before the war, Russian scientists were allowed to come to the United States. It is worth recalling that Soviet participation in the sixth Rochester conference was followed by the presence of fourteen Americans at a conference on particle physics held in Moscow in May 1956.

In spite of this success, however, we had not reached our goal, the repeal of the McCarran–Walter Act. Still, after the 1955 revision, the act was enforced more circumspectly and somewhat less radically. U.S. officials were quick to reproach others, like the Soviet Union, for their travel restrictions, even while curtailing travel into the United States by some foreigners. Happily, at the beginning of 1990, as the barriers came down in Eastern Europe, most of the McCarran–Walter restrictions were finally lifted.

Another occasion for personal contact with Soviet physicists occurred at the first international Atoms for Peace conference, which took place in Geneva in 1955. Taking advantage of the somewhat improved relationships between the West and the Soviet Union, the United Nations convened a conference on peaceful applications of atomic energy. Up to that time the two sides had worked independently and in the greatest secrecy on nuclear engineering.

Niels Bohr and many other important European and American scientists attended, along with a large contingent from the Soviet Union. It was the first time we were able to hold discussions about fission physics with our Russian colleagues. This was a subject that had been kept secret even as Soviet scientists were allowed to discuss other topics with Western scientists. We presented our findings, and they presented theirs. Until that time we had known nothing about their work. Their scientific papers contained the same measurements, the same curves, the identical numbers we had found in our research. This should not have been surprising, but still it made a deep impression on us. We had received an object lesson confirming that the supranational character of natural science transcends any national or ideological differences.

We discussed basic nuclear physics as well as its application to nuclear reactors for the production of energy. Medical applications—what we now call nuclear medicine—were also on the agenda. Any discussion of nuclear weapons was, of course, excluded. I was mainly involved in the sessions that dealt with basic science. Because my book *Theoretical Nuclear Physics,* written with John Blatt, had been translated into Russian, I was considered one of the main exponents of theoretical nuclear physics.

At the second conference, held in 1958 under United Nations auspices, the first exciting results coming from the small synchrocyclotron at CERN were reported. They dealt with the detailed mechanism of the decay of certain mesons.

During the two years preceding the 1958 conference I had found it strange not to have been invited to any of the meetings that had been held in the Soviet Union. I had so many friends there, and since arranging for the Soviets to come to Rochester, I had been recognized as an advocate of closer relations with the Soviets. Why hadn't I been invited to a single Soviet meeting?

At the 1958 Atoms for Peace meeting I asked Niels Bohr to look into this matter for me. He had a conversation with Vasily Emelyanov, who was an important man in the Soviet atomic enterprise. He told Bohr that the Soviet authorities were very dissatisfied with my report on my visit in 1955 and, in particular, with a remark I had made in the *Bulletin of Atomic Scientists.* I had ended my report by saying, "The reception of our Russian colleagues was warm and friendly almost to an excess in contrast to the attitude of the Soviet government." What I meant was that the government still saw itself in an adversarial position in relation to the United States. However, the Soviets took this very seriously because we had been invited by the government.

On my return from the USSR in 1955 I had passed through England and attended a private party at which a journalist was present. I told some of the jokes I had heard in the Soviet Union. Among them was this one: What is the difference between capitalism and socialism? The answer is: In capitalism man exploits man and in socialism it's just the other way around. The journalist published this and quoted the source. That also may have contributed to my exclusion from the Soviet Union.

When Emelyanov told him about these reproaches, Bohr was taken aback and tried to convince the Soviet that I was one of his country's staunchest friends, someone who was deeply interested in collaboration between the two nations. He said that this was all a misunderstanding and a big mistake. Because Emelyanov greatly respected Bohr, he allowed him to arrange a meeting between the three of us in which all was forgiven. Indeed, from that time on I was invited to Russian meetings almost every year.

Early in 1954, I received a letter from Oxford University offering me a professorship in theoretical physics. It was in the midst of the McCarthy period, and leaving the United States seemed more attractive than it had before. On the other hand, those times provided opportunities for me to fight against these dark forces. It was hard for me to make this weighty decision. Oxford made the offer very appealing by increasing my proposed salary through a permanent consulting agreement with Harwell Laboratory, the government nuclear physics laboratory nearby. I decided to accept on a trial basis and agreed to go to Oxford for two terms to see how I liked it.

The terms at Oxford were only three months long, so for my first term (April, May, and June of 1954) I went alone. I was a don in New College and lived in an extremely comfortable little apartment. I was provided with a manservant who attended to my every need. Except for a lack of central heating (I had only a fireplace), Oxford provided a most congenial atmosphere.

I ate breakfast in a communal eating room where by custom people were not to converse, because presumedly the early hour precluded interesting social conversation. Instead, each of us received a newspaper on a scaffold behind his coffee pot. The first day I was asked which paper I would like. When I inquired about the choices, I was told that "the gentlemen here prefer the *Manchester Guardian*." Since it was more liberal than the *London Times,* I preferred it, too. By contrast, in the evenings we were supposed to be sociable and speak to our colleagues, especially after dinner when we assembled in the Senior Common Room with sherry and discussed everything from world politics to Oxford University politics. Unfortunately, just like at the supper club in Boston, there was too great an emphasis on university politics for my taste.

When the chancellor or other visiting dignitaries came from London, special dinners were held at the so-called High Table in the big hall where all the students ate. Unlike the usual fare in British restaurants in those days, the food was excellent, because the colleges usually had very good French chefs. On one such occasion I was introduced to a strange Oxford custom. Great masses of food were served in course after course. Between the courses a fellow came around with a silver bowl filled with iced rosewater. He dipped a napkin into the bowl and moistened the area behind each person's ears. When I inquired about this, I was told it was supposed to rekindle the appetite. It seemed to have worked. I recall having eaten the following courses with great pleasure.

Oxford was a stimulating place, and I especially enjoyed my many contacts with colleagues in the humanities and other areas besides physics. Robert Nicholas, a historian, had been assigned to take me under his wing and teach me all the secrets of life at Oxford, such as when to wear a gown and when not to. We became very close friends in those days.

I was especially glad to be in the same community as Isaiah Berlin, whom I had met before and who had probably been involved in my invitation to Oxford. Berlin was a professor of literature who specialized in Russian and other Eastern works. He is one of the most stimulating men I have ever known. A great opera lover, he served on the board of Covent Garden Opera House, where we often went to performances and even to an occasional rehearsal. Many evenings we sat at the piano and sang Mozart operas together, after a fashion.

Isaiah married late in life. Aline, his wife, was divorced from my old friend Hans Halban, who had played an important role in atomic development in France and later in Canada. She was a member of the wealthy and influential Ginsburg family of Paris. When she and Isaiah married, he had to leave his apartment at Christ Church College, which was reserved for men only, and they moved to a rather princely residence in the Oxford area. Once I came back to Oxford very late from a trip and went to his house. We talked on into the night, as we often did. Around two o'clock in the morning I expressed a desire to eat something, and he asked me, "Where do rich people keep their food?" I thought we ought to try the refrigerator in the kitchen, where we indeed found something to eat. Apparently he had never been in his own kitchen.

Ellen had planned to join me in Oxford during my second term, in fall 1954. Since as a don I could not have my wife in the apartment, I had to rent a place in town. When I went to the warden of my house to say goodbye, I made a little joke about how times had changed, noting that dons once were not even allowed to marry. I continued that now at least they could do that, and perhaps one day they would also be allowed to live with their wives in the college. The warden seemed a little shocked and responded with great solemnity, "That will never happen."

Ellen was not impressed by the intellectual atmosphere at Oxford once she learned that women were relegated to a secondary role. She was incensed that they were not even allowed to eat with the dons. (I recently heard that things have changed. A woman can now be invited to the High Table, but she must be engaged in research or teaching. Wives are neither to be seen nor heard, as far as the college is concerned.) I was not surprised that this discrimination against women influenced Ellen's feelings about the place.

When it finally came time to make a decision about taking a more permanent appointment at Oxford, we decided against it, and I returned to my post as professor of physics at MIT. There were several reasons. First was the end of Senator McCarthy's influence in fall 1954. Also, most of our friends were in the United States and, more important, so were our children. Finally, we felt that the transition from being foreigners to becoming full-fledged members of the scientific community would be more difficult in England than it was in the United States. I have no regrets about our decision, especially because I feel that England's influence both in science and in the world as a whole has decreased since the time I was there.

I spent most of my second sabbatical in 1957 and 1958 at CERN. As was our custom during our European trips, we also went to Vienna, visiting old friends and recalling old memories. In 1957, I took my son, Tom, along when I visited the old director of my gymnasium, Professor Radnitzky. He had been a Social Democrat when he was director, and he had always supported my efforts in the party, including the cabaret performances that had taken me away from my studies. I was surprised that he recognized me immediately after thirty-two years.

During the Second World War, Radnitzky had spent some time in a concentration camp because of his political leanings. His daughter also had had a very difficult time. While still in high school, she had fallen in love with one of my colleagues, and they had had an illegitimate child. The father of the child was Jewish, whereas she was not. By the time the Nazis came, Radnitzky's daughter was married to another man who said the child was his in order to protect it from the Nazis. But the Nazis insisted on a blood test, which proved otherwise. They threatened to arrest her and her husband and take the child away. At the last minute, with great hardship, they managed to flee across the frontiers, over the mountains to Switzerland. I had had no idea of these events until Radnitzky told us this story. The visit with Radnitzky was very touching because he was so happy to see me and to meet Tom. By that time Tom had been at Heidelberg and could speak and understand German.

I visited another of my old teachers during that year, Professor Steppan, who had taught me mathematics in high school. I remembered him as a strict disciplinarian who ruled his classes with an iron fist, but he was an excellent teacher. Much of my mathematical education came from him, and he was able to transfer a certain enthusiasm for the subject. I had known that he had Nazi sympathies even at that time, but he did not show his anti-Semitism in class. Actually, I had felt that he liked me because he realized I had a gift for mathematics, and he singled me out for special help and attention in spite of my being Jewish. He had a great influence on me. I will never forget the day he taught DeMoivre's theorem, which deals with complex numbers. The theorem says that an exponential function with an imaginary exponent is actually a combination of cosines and sines. This may be unintelligible to some of my readers, but for me it was a startling surprise and such a fantastic revelation that I stayed up the whole night wondering about it.

Steppan had made no secret of his Nazi leanings. After the war, in the de-Nazification period, he had lost his job and was forced to take early retirement. By 1958 he was living with his sister in Salzburg. Since I owed him so much, I decided to go to Salzburg to visit him. He was probably about sixty-five years old but seemed a great deal older. I told him first how immensely indebted I was to him for what I had learned and how much it had helped me throughout my career. I also expressed gratitude that he had never allowed his anti-Semitism to come between us when he was teaching me. To my astonishment and dismay, he burst into tears and wept for a long time. I suppose no one had ever said anything like this to him.

During this sabbatical we also went to Israel to attend the ceremony at the Weizmann Institute commemorating the tenth anniversary of the death of Chaim Weizmann. (For a time I was a member of the board of directors of that institute.) Bohr and Oppenheimer also attended, and it was a very formal ceremony, almost Wagnerian in character, with music by the Israel Philharmonic and many speeches. A member of the Israeli air force, formerly an American, wanted to show Oppenheimer Israel from the air. Oppenheimer invited Ellen and me to join him on that flight. It was a fantastic experience. We flew all over the country, from the north to the south and over the desert and the West Bank, then occupied by Jordan. Especially interesting was the flight over the ancient fortress Masada and the Dead Sea. Then we flew south to Elath, where we swam in the Red Sea and went out in a glass-bottom boat for a wonderful view of the coral reefs and the exotic sea creatures.

Later we traveled by car through much of the country with the well-known Israeli physicist Yuval Ne'eman, who later became a reactionary right-wing politician. He showed us the country and told us many stories based on ancient and biblical history. According to Ne'eman, on Carmel Mountain near Haifa, in ancient times, there was a competition between the Baal cult and the Jewish religions. The two rival priests asked their respective gods to set some dry wood on fire. The one whose god obliged would be said to have the stronger religion. The Baal priest could not get the wood to burn, but the Jewish priest succeeded. Ne'eman told us that the Jew had probably had a little lens hidden in his hand and had cheated by using science to win what was supposed to be a strictly religious competition. True or not, this story certainly reveals Ne'eman's pro-science bias.

My visits to Israel, beginning with the first one in 1956, strongly influenced my feelings about my Jewish origins. As a teenager, I was aware that my family was Jewish, but it did not mean very much to me. At most, perhaps I felt some pride about belonging to a group that emphasized intellectual values. I never experienced any acts of anti-Semitism directed at me personally, except for a few clashes with rightist youths when I was a member of the Socialist youth group. Inasmuch as our upbringing was not religious, I had never gone to a synagogue until I visited Israel, I had had no bar mitzvah, and we had never celebrated any Jewish holidays. My brother and I were not even circumcised. I first attended a Passover Seder in the 1970s in Ann Arbor when I visited my son (who is also agnostic), and we were invited to the home of one of his more religious friends.

Of course, the rise of anti-Semitism through the Nazi movement had made the Jewish question more acute. However, I had reacted against these inhuman excesses not by feeling more Jewish—I was too much of an internationalist. The international character of the scientific community and also of the Socialist movement made me feel that I was a world citizen who abhorred all nationalist tendencies. I expected that humankind would overcome the barbaric cleavages between nations, races, and religions through a growing aware-ness of humanity's common interests. I felt cultural and racial differences should have positive effects, adding to the variety and cross-fertilization of human cultures instead of causing conflicts and destruction. These optimistic visions may have been the effect of my upbringing in the Jewish community in Vienna, a community bent on assimilation. That community was swept away during the Nazi period.

In the 1970s, when I was offered the directorship of the Weizmann Institute of Science in Rehovot, it was clear to me that I had to refuse this offer. Earlier when I had enthusiastically accepted the director-ship of CERN, it had been because of its inter-European, interna-tional character. I believe that the nationalistic solution of the Jewish problem in Israel cannot be the answer for all Jews, and this view affected my assessment of the Weizmann Institute offer.

Nevertheless, my first visit to Israel impressed me greatly, partly because of the astounding success of the early Zionists who built a functioning country with industry, science, and agriculture virtually from scratch. I was equally impressed by the sight of Jewish police officers, a phenomenon that seemed almost a contradiction in terms to a person who had grown up in Vienna.

Another experience shook me deeply. While in Israel, I received a telephone call from a classmate from my Vienna gymnasium who had immigrated to Israel. He was from a poor Jewish family and had always lived in fear of being attacked by anti-Semites. His behavior in Vienna had been characterized by a sense of inferiority, a lack of self-confidence that grew out of this fear. He had walked with his head bowed. When I met him in Israel, he had become a completely different person. He held his head up and exuded a sense of self-confidence and pride. He told me about his important job in the foreign office and talked about the great future that he saw for Israel.

During my first visit to a synagogue, I was impressed by the holy character of the scroll, which the rabbi took out of a golden chest during the service. It was not a picture or a statue, not a religious relic; it was a book, the symbol of learning and scholarship. I was reminded of the tearful embrace of Bob Marshak's father as he said, "My whole life with all its deprivations were worthwhile; my son is now a professor."

These experiences made me more conscious of the significance of my being Jewish. I consider it a privilege to belong to a heritage that highly values things cultural and intellectual. But I believe that such a tradition is not the province of one group of people. It is valid all over the world, wherever cultural values are created.

In September 1958 we returned to Cambridge after spending the summer in Altaussee. Late that fall Pauli died of cancer. He was only fifty-eight years old—as old as the century. I went to Zurich in December to take part in his memorial service, which was held in a church, although he had not been a religious man. The stained-glass windows by Chagall, for which the church was famous, gave the services a special beauty. From then on my relationship to Mrs. Pauli was very close. She considered me the executor of Pauli's intellectual estate, and later I helped edit Pauli's works and his letters.

The most important postwar political problem was (and still is) the increasing danger the nuclear arms race posed to the world. Many of my colleagues and I actively continued trying to explain this problem to a very poorly informed public. Most polls continued to show that Americans were convinced that we were better off because we had the nuclear bomb and that the Soviets would never be able to match our power with these weapons. Many of us gave speeches and took part in public information programs designed to educate the public about the terrible catastrophe that could annihilate us all.

Leo Szilard, who was especially active in these endeavors, proposed that we get together a group of well-known scientists who would devote themselves to this task. The Emergency Committee of Atomic Scientists was founded in 1946, and Szilard convinced Albert Einstein to serve as chairman. Einstein was certainly the best-known scientist in the world, and his chairmanship had an impact on the public. There were eight members of the committee: Albert Einstein, Leo Szilard, Hans Bethe, Thorfin Hogness, Harold Urey, Philip Morse, Linus Pauling, and myself. The funding came from many sources, and we were able to raise several hundred thousand dollars. We met several times and produced statements, usually composed by Szilard. The press reported these statements, but I don't think we had any important influence on public opinion. Altogether it took about forty years for people really to wake up to the threat of a nuclear catastrophe. It was only in the late seventies and eighties that the idea of the absolute impossibility of "winning" a nuclear war was recognized not only by the public but also by governments.

One of those especially eager to help the scientists in their efforts for peace was the wealthy Cleveland, Ohio, industrialist Cyrus Eaton. He was born in Canada and had an estate in the village of Pugwash, near Halifax, Nova Scotia. Eaton asked Szilard how he could be useful. Szilard proposed an international scientists' organization devoted to the prevention of nuclear war and to arms control in which scientists from both Western and Eastern Europe, the

United States, and the Soviet Union could meet to discuss these matters. With Eaton's support, Szilard organized the first of the Pugwash conferences in 1957. Eaton came to the conference in his private railroad car lavishly outfitted with bedrooms and a study. A number of Soviet scientists also attended. This was the first occasion on which not only science but the political problems of nuclear disarmament were discussed in a group that included Soviets.

In subsequent years Pugwash conferences were held at many different places in the United States, Europe, and the Soviet Union. There were meetings devoted to special problems and plenary meetings dealing with the general question of world peace. The attendance was always international, with good representation from the Communist countries. The Pugwash movement still exists today. As an aside, at the 1958 Pugwash meeting in Kitzbühel, Austria, a celebration was organized in honor of my fiftieth birthday. It had been planned without my knowledge and was a total surprise. I was very moved to be so honored in the country of my birth by my colleagues from both the East and the West.

Pugwash more than any other organization embodies Bohr's dream of the internationalization of nuclear matters. I owe my participation in this exciting venture to Leo Szilard, who dragged me into these things even when I thought I didn't have the time. I am grateful that he repeatedly convinced me to participate. Through Pugwash we had a rather direct line of communication with the Soviet government. The Western participants attended these meetings without any official ties to their governments. Indeed, some of them vigorously opposed their countries' cold war policy. In contrast, the Communist participants could join the meetings only if they were approved by and closely connected to their governments. For once, the USSR's restrictive policies worked in our favor by putting us in contact with very highly placed people.

Two important ideas came out of the Pugwash discussions. One was the urgency of finding a way to end the testing of nuclear bombs. This was partially realized when tests in the atmosphere were prohibited. The success of this effort was due in no small measure to the efforts of Linus Pauling, the famous American chemist. He convinced the public that nuclear bomb tests in the atmosphere spread dangerous radioactivity over the earth's surface. He received the Nobel Prize in chemistry in 1954 and the Nobel Peace Prize in 1962 for his success in stopping bomb tests in the open. That partial halting of tests prevented further radioactive pollution, but it was not very effective in putting an end to the development of new nuclear weapons. Both the superpowers continued testing underground. Only a comprehensive test-ban treaty would slow down the unnecessary and dangerous increases in the power of existing nuclear weapons.

The other important contribution of Pugwash was the idea of an antiballistic missile (ABM) treaty, which was finally concluded as part of the so-called SALT I treaty (for "strategic arms limitations talks"). By restricting the construction of devices that destroy incoming missiles, the ABM treaty prevented the superpowers from defending themselves against ballistic missiles. The idea was that the development of defensive weapons would only force the other side to stockpile more missiles, thus accelerating the arms race. At first the Soviets strongly opposed this idea. They said, "What is wrong with defensive measures? Why should we be against defense?" It took more than two years to convince the Soviet scientists of the logic of the ABM treaty.

At that time the U.S. government understood the ABM problem very well. The Americans were actually ready to sign the SALT I agreement, including an ABM treaty. Only the Soviets needed to be convinced. Thirty years later, in the 1980s, the situation completely reversed; the American government tried to subvert the ABM treaty by developing space defenses against incoming missiles, a task that was probably technically impossible. This so-called strategic defense initiative (SDI), or Star Wars program, would be a violation of the ABM treaty.

Pugwash served as a useful bridge between the East and the West on ideas about arms control. There was, however, an asymmetry in our roles, since the Eastern participants always reflected the views of their governments, whereas Western scientists were free to express their own opinions. This made it very tricky when we got around to composing the final statements after the meetings. I was often involved in negotiations in which we tried to get both sides to see the other's point of view. Because of my European connections and my many visits to the Soviet Union, I was able to help in this endeavor.

Pugwash encouraged the birth of other organizations with similar aims, and some of them continued what Pugwash had begun. For instance, the National Academy of Sciences in America has a common committee with the Soviet Academy of Sciences in Moscow where political questions of mutual interest are discussed. Pugwash has lost its unique role in these matters in the last decades, but it remains an important organization that will continue to be called on in the future.

# ELEVEN

⌗

# Leading CERN, a European International Laboratory

FROM THE END OF the Second World War up to 1960, my main professional interest had been nuclear physics. With the publication in 1952 of my book *Theoretical Nuclear Physics,* I found myself deeply involved in all aspects of the field. I was invited to most of the meetings on that subject, where I was asked to give introductory and summary talks. I also served on committees for planning and research. In the 1960s, my interest in nuclear physics remained strong, but I also turned my attention increasingly to the field of particle physics.

Whereas nuclear physics was mainly concerned with how neutrons and protons (which are jointly referred to as "nucleons") combine to form nuclei, particle physics concentrates on the internal structure of the nucleons themselves and studies the many new short-lived particles that appear when the nuclear constituents or electrons collide at very high energies. For this reason the field is often referred to as "high-energy physics." Particle physics goes more deeply into the structure of matter than nuclear physics. However, the phenomena investigated are further removed from our natural environment, since the very high energies necessary to produce them are rarely available in nature. We create such conditions by using powerful accelerators that produce beams of accelerated particles. The particles hit a target or collide with each other. Such conditions probably occurred in nature at the origin of the universe and are present now only when cosmic rays penetrate our atmosphere. Cosmology, the study of the beginning of the universe, and particle physics, the study of the basic components of matter, come together here.

At the early stages of the universe, shortly after the so-called big bang, temperatures were enormously high. Matter was not composed as it is now, but was dissociated into its most elementary constituents. At the targets of our accelerators, each time the energetic particles hit other particles, we are reproducing, more or less, conditions that existed when the universe began. I confess that, as a physicist, I feel a certain pride that we were able to create such cosmic conditions at the targets of our accelerators.

Until 1960, I never had enough time to devote myself to a thorough study of these problems, although I had always been interested in particle physics. I sometimes gave survey talks about the subject, but more often I was a passive participant at talks and discussions, such as the Rochester conferences, that dealt mostly with particle physics. Nuclear physics expanded rapidly during the fifties and sixties, and several of my colleagues asked me to write a second edition of our book, but I did not think it would be an interesting project. Most of the new developments that would have been in such an edition were only extensions and continuations of previous work. For some time I had been feeling anxious to go on to something new, but there were so many other demands on my time that I found it hard to make such a change.

The event that provided me with that opportunity was the tragic death of the director general of CERN, Cornelius Bakker, in an airplane accident in April 1960 as he was arriving at Idlewild Airport for an American Physical Society meeting, where he was to report on the progress of CERN. When I heard of his death, as sad as the event was, I felt immediately that this might be a turning point in my life. And that turned out to be the case. But before I talk about my role in it, I should give a bit of background about CERN and how it came into being.

Before the Second World War, Europe was the cradle of science, especially fundamental physics. As mentioned earlier, the most important developments, such as relativity theory and quantum mechanics, were conceived and developed in Europe. But after the war, America took over. A wave of scientific inquiry swept over the

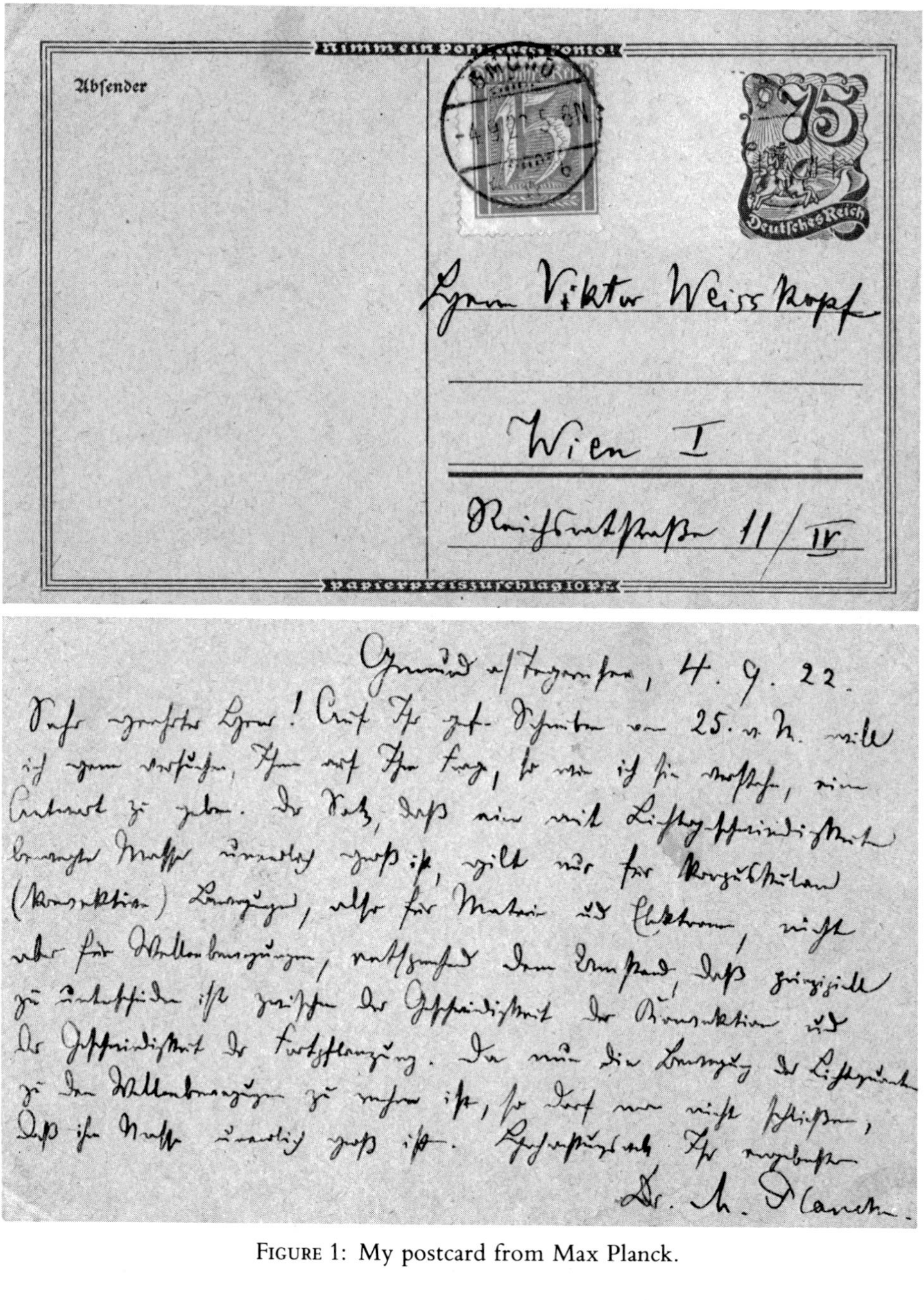


FIGURE 1: My postcard from Max Planck.

FIGURE 2: I was four and Walter was two in this picture taken on the beach in Belgium with our parents.

FIGURE 4: Tante Toni creating a waterfall of sound on her piano.

FIGURE 3: Our father's beard scared us when he returned from the war.

FIGURE 5: My social conscience came from Uncle Carl.

FIGURE 7: On the way to the Institute in Göttingen by bike. (L. to r.: Myself, Misi (Maria) Göppert, Max Born.)

FIGURE 6: Die Urmama, the founder and guiding spirit of our family.

FIGURE 8: My friend, the great Russian scientist Lev Landau.

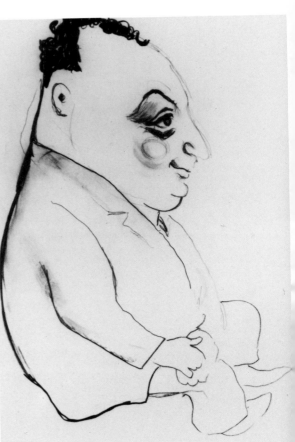

FIGURE 10: E. Rabinowitch's pencil sketch captures Wolfgang Pauli better than any photo could.

FIGURE 9: Paul Ehrenfest taught me that physics is "simple but subtle."

FIGURE 13: Eugene Wigner, my mentor for my Ph.D., talking to Henry Barschall.

FIGURE 12: Erwin Schrödinger in a typical pose.

FIGURE 14: A wonderful sculpture of Max Planck that was rejected by the Communists of East Berlin now stands in the courtyard of the High Energy Laboratory in Zeuthen.

FIGURE 15: An idea is born: (from l. to r.) John Blatt, Bruce French, Herman Feshbach, and myself as young associates at MIT, *circa* 1949.

FIGURE 16: Hans Bethe may be telling me by phone that my calculations were wrong again.

FIGURE 17: Werner Heisenberg and I discuss the future of CERN.

FIGURE 18: Ellen was a dancer all her life. Here we are at my retirement party at MIT.

FIGURE 19: Pope John Paul II, Paul Dirac, and I at the Einstein memorial at the Vatican.

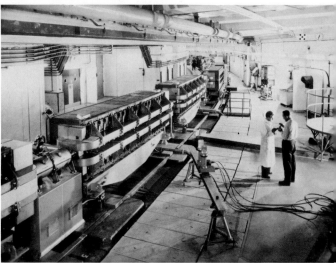

FIGURE 20: Inside the big accelerator at CERN. At left, the magnets that bend the proton beams around the circular track.

country, aided by the influx of refugee scientists from Europe. While Europe's economy was still recovering from the ravages of war, money was readily available in the United States for the expensive accelerators and detector systems that nuclear and particle physics require. Some of these accelerators were so costly that it was difficult for any European country to afford them, even after the recovery from the consequences of the war had been accomplished. In the late forties, therefore, European scientists were already discussing the possibility of some kind of joint effort, a pooling of financial and intellectual efforts that would facilitate the rebirth of European science.

Pierre Auger, who was at that time responsible for science at UNESCO (the United Nations Educational, Scientific, and Cultural Organization), had been involved in these plans along with many other members of the inner circle of the world physics community. Among these were the administrator of the French Atomic Energy Commission, Raoul Dautry, and Edoardo Amaldi, a well-known and influential physicist at the University of Rome who became my close friend. In December 1949 a European cultural conference in Lausanne, Switzerland, established a European Cultural Center to promote common cultural activities in Europe. On that occasion Dautry read a message from the famous French physicist Louis de Broglie proposing an international research effort in which all European countries should participate.

When people discussed the possibility of an international project in physics, they turned quite logically to particle physics rather than to nuclear and solid state physics. The latter fields were already deeply involved in industrial applications. Furthermore, since nuclear physics was connected with weapons, it would not have been appropriate for an international organization. It also made economic sense for countries to work together on particle physics projects because they required large and expensive equipment. Finally, basic research into this relatively new area would continue the European tradition of searching for the elementary constituents of matter, which had begun with Newton and had culminated in the explanation of atomic and nuclear properties through quantum mechanics.

Interestingly, however, it was an American, Isidor Rabi, who gave the project its first push. At the general conference of UNESCO in

Florence in June 1950, he proposed that the European nations should get together to start a common laboratory, emphasizing that it was important for the Europeans to join efforts in fundamental physics, which at that time were mostly in the hands of Americans. This was a very significant step, scientifically as well as politically. I have been told that, at the time, some of the American delegates were not so sanguine about a proposal to promote competition with America. The weight of Rabi's personality and reputation got the idea through, however, and it was accepted enthusiastically in Europe, especially by those who were already considering similar ideas.

During the same year, at the meeting of the European Cultural Center, a commission on scientific coooperation had been appointed under the chairmanship of the Swiss writer and philosopher Denis de Rougemont. Small amounts of money were coming in to support the work. In all, about $10,000 had been collected from several countries. It was by then an accepted notion that an international institution devoted to particle physics and using the most up-to-date equipment would be the best solution. The plan was to build a very large accelerator, the biggest one possible at that time, as well as a smaller one, which would help the scientists gain experience in operating such devices. The project was named CERN (Conseil Européen pour la Recherche Nucléaire), or the European Center for Nuclear Research, a slight misnomer because it was *not* going to be a nuclear physics research lab, but one devoted to particle physics. A provisional organization was established in May 1951, and a board of consultants appointed Edoardo Amaldi as secretary. His early enthusiasm for the idea of CERN and his inspiring personality made him the obvious choice. Various Western European countries contributed a few hundred thousand dollars in new funding for the project.

Today such international efforts take a long time because they demand a lot of bureaucratic preparations. CERN is notable for the extreme speed with which it was developed. By the beginning of 1952 four study groups had been organized, each with a specific mandate. Cornelius Bakker's committee was to design the smaller machine, a synchrocyclotron for accelerating protons to 600 million electron volts (MeV). A second group, headed by the Norwegian engineer Odd Dahl, planned a proton-synchrotron that would accelerate protons to at least 10 GeV. (The abbreviation for billions of electron volts was BeV in the United States. In Europe the expression GeV, standing for giga-electron volts, was used since the word *billion* has a different meaning there.) Lew Kowarski, a Russian-born French citizen, led a third group, which was to plan the construction of the laboratory and the establishment of the organization's infrastructure. The fourth group, the theory group, under Niels Bohr's leadership, was initially located in Copenhagen. Bohr had not been an early supporter of CERN. Perhaps he was afraid that it would compete with his Copenhagen institute as an international organization. But later he was among CERN's most active supporters and was among those who urged me most strongly to become its director.

When the time came to decide where to locate CERN, there were a number of proposals. Copenhagen was mentioned, as was Holland. But finally Geneva was chosen, partly because it was the headquarters of so many other international organizations and therefore had many facilities, schools, and clubs for foreign residents. It is a testament to the foresight of CERN's founders that they chose a site adjacent to the Swiss–French frontier. This prescient choice of location later permitted the laboratory to become truly international when some important additions were constructed on the French side.

The members of the provisional organization received low salaries or worked only part-time, because there wasn't much money available for them. A number of American physicists and engineers with experience in the construction of the necessary machines served as advisors and consultants for the group. Most Americans had no prejudice against building a competing laboratory. On the contrary, the idea of an international institution seemed to attract U.S. scientists who were eager to offer the benefits of their greater experience and to share their know-how.

During the early fifties a new "strong focusing" principle for acceleration was invented in the United States at Brookhaven and other places. This principle made it possible to reach much higher energies with a machine of a given size and cost than had been possible with older methods of acceleration. Luckily, this new method was developed before the European group had made a final decision about what kind of machine to build. With the effort and money they would have spent to reach an energy potential of ten billion electron volts, they could now reach almost thirty billion electron volts.

The legal and organizational work required to establish the permanent organization was completed by January 1953, and on 29 June in Paris the provisional organization was transformed into a permanent one, the last step before ratification by the participating governments. This transformation made CERN a legal reality. Financial support increased dramatically. During the interim period, the budget had grown to about three to four million Swiss francs, which at that time corresponded to about a million dollars.

Finally, CERN became a working reality in 1954, only four years after Rabi had first proposed the idea. An architectural firm had drawn up the plans for the layout of the laboratory, and construction began in May of that year at the present site in Geneva, near the French frontier, facing France. In September 1954 the project was ratified by the participating nations: Belgium, Denmark, the Federal Republic of Germany (West Germany), France, Great Britain, Greece, Italy, the Netherlands, Norway, Sweden, Switzerland, and Yugoslavia. Austria and Spain joined later. The East European nations were invited, but Stalin did not permit their participation. Only Yugoslavia accepted, because Marshal Tito, the country's leader, saw his acceptance as an act of defiance against Soviet domination. (However, in 1961 the Yugoslavs dropped out for lack of funds.) CERN's rapid transition from idea to actuality constituted a major triumph over the bureaucratic inertia that usually paralyzes such massive international projects.

The birth of CERN came at a time of great enthusiasm for European cooperation in general. An early supporter of European unity was Jean Monnet, one of the architects of the European Common Market, who with a handful of others recognized that the European countries, despite all their differences, were part of a common culture and tradition and had similar political aims. The vivid memory of a devastating inter-European war inspired their determination that such a tragedy should never occur again. The times demanded that they band together in common causes—cultural, scientific, and economic. In addition to the idea of a united Europe, the other driving force was the resurgence of science in the United States, which fired up the ambitions of the European scientific community for a similar European scientific renaissance and made them eager to get CERN established. Most of the scientists at CERN in those early days were relatively young, in their thirties or early forties. Apart from Edoardo Amaldi, there was Gilberto Bernardini from Italy, Guy von Dardel from Sweden, Peter Preiswerk from Switzerland, John Adams and Mervyn Hine from England, Wolfgang Gentner and Anselm Citron from West Germany, and many others whose enthusiasm energized CERN.

I have always thought one of CERN's strengths has been its independence from any other international organization. Preliminary discussions about CERN began at UNESCO, but UNESCO was (and is) a world organization. CERN, however, was perceived as a European affair and soon established its own identity. The organization was set up so that the highest authority resided in a council consisting of two representatives from each of the member states. One of these representatives was always a scientist, the other an administrator.

The extent of each country's financial contribution was based on its gross national product. For example, France, England, and West Germany each contributed roughly a quarter of the CERN budget, and the smaller, less affluent nations paid the balance. A very wise rule prohibited any country from paying more than a quarter of the budget. Later, West Germany became so prosperous that, according to a proportionate calculation, it should have paid over 30 percent. But the rule remained in force and served a very important function, since it prevented any one country from exerting too much influence.

In the early days of CERN the memories of the Nazi past were still strong, but it was clear from the start that for the sake of a truly united effort, Germany had to be accepted as an equal partner. Isolating Germany might have led to another nationalistic explosion. As CERN director, I frequently traveled to Germany, and I made it a point to mention that country's terrible past, not so much as a reproach but as a challenge for a different Germany. I saw scientific collaboration as an important lever for integrating Germany into a united Europe, which would strengthen its newly developing democratic spirit.

There were two prominent German scientists, Werner Heisenberg and Wolfgang Gentner, who in their own different ways contributed to the development of CERN. Their work during the Nazi time was described earlier. After the war, Heisenberg devoted much of his time and effort to rebuilding science in Germany and reestablishing the bonds between German science and the rest of the world. (After all, he had experienced the glory of international science during his youth at Bohr's institute in Copenhagen.) He encountered some difficulties, however, because of the widespread lingering distrust of Germans and a residue of doubts about his role.

Heisenberg was also disappointed that his last attempts to formulate an all-embracing theory of matter did not succeed. He thought that he had found the fundamental law that explained the world of elementary particles, but he did not convince his colleagues of his ideas, which ultimately proved inadequate. It was one of the few times that Heisenberg's intuition led him astray. He was so sure of his new theory that he was less than enthusiastic about constructing more penetrating accelerators at CERN, since he believed his ideas answered the questions we were studying. But in spite of his state of mind, he was very helpful in getting CERN going and in getting Germany to play a constructive role in that enterprise.

Wolfgang Gentner, a very different character, was extremely active in the organization of CERN. He spent some time in Geneva so he could assist in the construction of the laboratory, and he was ready to help whenever his counsel was needed. His energy and wisdom were of the utmost importance to the success of CERN. His anti-Nazi activities during the war made him an ideal representative of a new and better Germany in the family of European nations. He remained skeptical about Heisenberg—in my view perhaps too skeptical, since he distrusted not only Heisenberg's politics but his scientific judgment as well. When CERN was established, the subject of involving German scientists still occupied much discussion. Nevertheless, these two Germans both contributed to CERN's early days even though they were very different from one another in their wartime experiences.

CERN's first director general was my close friend, the Swiss-born American Felix Bloch. He was an outstanding physicist, a Nobel Prize winner, the founder of the theory of metals, and one of the inventors of nuclear magnetic resonance (NMR), with its important medical applications. He was one of the great physicists of this century, both as a theorist and an experimentalist.

Before he accepted the job, Bloch asked me whether he should do it. I told him, "Felix, this is not your style. You are a physicist, and you are not interested in management. The first years of CERN will consist of organization, engineering, and the construction of buildings, machines, and laboratories. You are interested in physics, especially in your own experiments. I don't think this is the job for you." He accepted anyway. One condition of his acceptance was the transfer of part of his laboratory from Stanford to Geneva so that he could continue his research, which was unquestionably his primary concern. Indeed, from the beginning difficulties surfaced. Bloch's interest in CERN was often diverted by his preoccupation with his own research, and Edoardo Amaldi had to take over many of his duties. The unworkability of his appointment became obvious. Bloch gave it up after one year and had his equipment moved back to California.

Cornelius Bakker from Holland succeeded Bloch. He was older than the other candidates and had a lot of administrative experience. (I still think Wolfgang Gentner would have been a better choice, but the fact that he was German probably worked against his appointment at that time.) When Bakker took over, CERN was nothing more than a big construction site, muddy when it rained and dusty when the sun shone. The different parts of the laboratory grew up like mushrooms here and there. In what seemed like a matter of a few weeks the housing for the small accelerator and the tunnel for the big proton-synchrotron emerged, as well as workshops, office buildings, and cafeterias. Odd Dahl, who was in charge of the construction of the proton-synchrotron, had to return to his native Norway, and his deputy, Frank Goward, died suddenly of a brain tumor. Somebody else had to be found to take charge of this aspect of the project.

England's John Adams was chosen. He was a self-made man who not only had no Ph.D. but no university training. However, he was a first-rate engineer and organizer, full of excellent technical ideas and in many ways a great man. Thanks to him the proton-synchrotron is still one of the best machines of its type. By the time the small machine was finished in 1957, some physics research had already begun under the leadership of Gilberto Bernardini.

When I became director in 1961, I asked Bernardini to stay on as director of research, a position he had held for several years. He was a great character—very volatile, impulsive, and enthusiastic. He insisted that he was not Italian but Etruscan, and indeed he looked like a figure from an Etruscan fresco. Occasionally he was difficult to work with, since hardly anyone could keep up with his many, ever-changing ideas. But it was always a pleasure to listen to him, and we discussed not only physics but also music, art, and philosophy.

One of the first experiments performed with the small accelerator resulted in an important discovery. The experiment was performed by Giuseppe Fidecaro, Alexander Merrison, and Alvin Tollestrup and dealt with the decay of the pi-meson, the short-lived particle that appears when protons of high energy collide with matter. The pi-mesons decay mostly into muons, but the theory requires that a few should also decay into ordinary electrons. Such decays had been missed before, which made people doubt the reliability of the theory. Theorists felt exuberant when the group at CERN was able to observe such decays. Fidecaro announced the news of the discovery in a report to the international Atoms for Peace conference in Geneva in 1958. Richard Feynman performed a joyful dance on the floor of the conference hall when he heard about it.

I followed the development of CERN with great interest and enthusiasm, and I spent my sabbatical during the 1957–58 academic year there. For me the idea of an international scientific laboratory was a symbol of the European unity for which I longed. The big machine produced its first accelerated proton beam—but not yet in full force—at the end of 1959, five years after its construction. At the meeting of the American Physical Society in fall 1959, I could announce that the proton–synchrotron, the first accelerator based on the new strong focusing principle, was working well.

While I was at CERN in 1958, Bakker asked me to become his director of research. I didn't accept his offer at that time because we had different ideas about how CERN should be run. When I heard of his death and it occurred to me that I might be offered the job again, I was ready to accept because it would provide me the opportunity I had been looking for to delve seriously into particle physics.

Immediately after Bakker's death, I did receive a number of calls asking me to come to Europe to discuss the possibility of some kind of leading position at CERN. John Adams, who was acting director after Bakker's death, was eager for me to come as scientific director. But then the British government suddenly asked him to return to England to start a fusion laboratory in Culham. It was an offer he could not refuse.

For a while it was unclear how the leadership of CERN would develop. I talked to many people in the United States and in Europe about what I should do if I was offered a leading job at CERN. Some Americans advised me not to touch a job at CERN because, they said, the Europeans would never be able to run a big laboratory, and the job would be difficult, troublesome, and ultimately thankless. I did not feel that way. Although I expected great difficulties, I was convinced that the European scientific tradition was strong enough to overcome them. In addition to being attracted by the international character of CERN, I had a special, personal reason for wanting to take the position. I wanted to be involved directly in the rebirth of European science, as a symbolic act of gratitude for all the cultural inspiration the continent had given me. And on another level, I confess that the wonderful surroundings of Geneva, with its excellent opportunities for skiing, were an additional enticement.

One of my students, Hans Mark, encouraged me to take the job of director general. He had an excellent reading of the situation. "In a secondary position," he said, "you will have to convince the people below you as well as the director general of your ideas and projects. If you take the top job, you'll be in a position of power and will be able to do what you think is right." It was very good advice, and when John Adams offered in December 1960 to propose me for the job of director general, I accepted.

Up until that point in my life I had worked with graduate students and a handful of older collaborators. But I had never had the responsibility of dealing with large, varied groups of people that included not only scientists but administrators, engineers, craftsmen, politicians, and statesmen—unless you count my chairmanship of the town council of Los Alamos, which I considered a great honor even though it wasn't much of a job. The job at CERN promised me an opportunity to broaden my experience and try out my abilities in this area.

The formal offer came from the Greek member of the governing council. Then the council chairman, François de Rose, personally offered me the position. De Rose had been involved in the idea of CERN for a long time. He is a wonderful man, extremely conscientious and well informed in science and in the humanities. He is also a high-ranking French diplomat. In short, he is one of those well-rounded intellectuals France has produced. I was invited to come to Paris to be presented to the council. The council members were extremely friendly and eager for me to accept the post, but they had a few questions. One of them asked whether I had had any administrative experience. Without hesitation I answered, "None whatsoever, but I consider this my strength." I said it on impulse, but I found out later that there was a good deal of truth in it.

Taking the job meant leaving MIT and the Cambridge community of which we were so fond, but I only planned to be at CERN for two years under a leave of absence from MIT. As it turned out, I stayed for almost five. The actual date of my appointment was 1 July 1961, but I decided to go to Geneva at the beginning of that year in order to get acquainted with CERN's day-to-day operations. Our children were already in college, so it was relatively easy to make arrangements. We left for Switzerland, via New York, early in February, but were caught in one of the heaviest snowstorms New York had ever experienced and were stranded there for three days before the airport reopened.

After this inauspicious beginning we set up our life in Geneva, but a few weeks later something else happened. On 13 February, I was being driven home by Georges Dakin, CERN's director of admin-istration. It was a foggy day, and Dakin hit the back of a truck. Dakin was not hurt, but I was pressed between the dashboard and the seat, and my right hip was gravely injured. At the hospital in Geneva they didn't know what to do for me. The acetabulum, the hollow within which the head of the femur moves, had shattered into fifteen pieces. A group of European hip specialists was called in, but all they could suggest was fusing the hip, with the result that the joint would remain stiff. I was asked whether I wanted it to be fixed into a sitting or a standing position. I accepted neither choice. I was in terrible pain, but I knew there had to be another solution.

I sent my X rays to the Massachusetts General Hospital (MGH) in Boston, because I knew of their excellent care. I had met many of the MGH doctors when I served on their scientific advisory board as the one nonmedical person appointed each year to work with the physicians. Dr. Otto Aufranc, a leading hip specialist and a pioneer in hip reconstruction who was at MGH, sent a cable telling me to come back to Boston immediately. He had reviewed the X rays and thought there was a possibility that the hip could be reestablished so that I could function normally again.

I have never had a more miserable transatlantic crossing. Propeller aircraft were still in use in those days, and there was no direct flight to Boston. Two rows of seats had to be removed from the plane so that I could be accommodated on a stretcher. I spent the seemingly endless trip in an uncomfortable position, and when we got to Paris, we found we had missed the connection from Paris to Boston. This meant several miserable, painful hours at the airport and a resched-uling via New York. When we arrived in New York, there was no immediate plane to Boston, and again I had to wait. After what seemed like a week of takeoffs and landings, we finally got to the hospital and Dr. Aufranc's examining room.

He told me there was a 50 percent chance that he would be able to repair my injury. What I was most eager to know was whether I would be able to ski again, since I would be in Switzerland and couldn't bear the idea that I would be unable to ski while living there. Aufranc must have thought I should be grateful he was giving me a chance just to walk again. No, he didn't think I would ever be able to ski again. I asked him if he was a skier, and when he said no, I decided not to believe him about my chances. And indeed, two years after the operation I was able to ski. From then on I sent Aufranc a postcard from every ski trip I took.

During the lengthy operation a metal cup was placed over the head of the femur so that the acetabulum could grow together without hindering the motion of the head. This was still a new procedure in those days, and Aufranc wrote a paper about my particular operation. (Today, in an operation that has become commonplace, an artificial joint is inserted.) I had to stay in the hospital for more than three months. From the first postoperative day I did exercises under the supervision of a carefully trained nurse, and when I went back to Geneva, I was able to get around fairly well, first on crutches and then without them. Slowly I began to walk in something approximating my normal gait.

Some years later I got an honorary degree from Oxford and was introduced to the famous English orthopedist José Truetta. I told him that we had already met—when he was one of the doctors called in to see my hip after my accident. He was astounded that I was able to walk and that my hip seemed to be functioning normally. I explained that Aufranc had put a metal cup over the femur so that the acetabulum grew together into one unit. Truetta was flabbergasted and insisted that I come to his hospital the next day so he could x-ray my moving leg and show his colleagues what Aufranc had accomplished. He later told me that he had tried that operation 200 times and had never succeeded. When I told this to Aufranc, he said that the secret was in the extended postoperative care, which the Europeans were not accustomed to giving their patients. Thanks to Aufranc, I have had more than twenty normal years complete with skiing, hiking, and mountain climbing. I will always be grateful to him.

Because of the accident I was unable to use the first half of 1961 to acquaint myself with the day-to-day business of CERN, as I had planned. Instead, for almost four months, until I returned to take over my job in July, I was out of touch with the people at CERN. I was not completely idle at the hospital, however. I wrote *Knowledge and Wonder,* a popular introduction to present-day science. I was proud of the last sentence: "Nature, in the form of man, begins to recognize itself." The spirit of the book is characterized by a Goethe poem, translated by my son-in-law, Douglas Worth, which I used as an epilogue:

> Joyfully a patient lover
> Mankind's eager spirit strove
> To unravel and discover
> How creative Nature wove.
> It revealed that one eternal
> Essence permeated all
> In the grains of shell and kernel
> At the heart of great and small.
> Always changing and yet holding
> Things together far and near
> Shaping endlessly unfolding
> To behold it we are here

Finally, in June 1961, I was discharged from the hospital, and in July I arrived at CERN to begin my duties.

When I became the director general of CERN, the laboratory had only existed for five years, which were mostly taken up with construction. Only a few experiments had been performed. My first task was to establish a sensible research program. As I began my tenure, I saw that a vital and ongoing part of my job would be to encourage the organization to reflect the two aims of CERN: first, to be an outstanding research institution imbued with the traditional spirit of the quest for scientific truth and, second, to be a successful example of European collaboration.

When John Adams became acting director general, he initiated a reorganization of CERN into a number of divisions, each in charge of an aspect of the laboratory, and a small directorate to assist the director general in coordinating the divisions' efforts. There were two divisions devoted to running the two accelerators and one for building detectors, as well as divisions for the different types of experiments, for heavy-equipment construction, for computerized data collecting, for theoretical physics, and even for housekeeping. It was a little awkward for me to accept the new form of organization, which had been established without my agreement. But I do not consider organizational charts of primary importance; what counts is the spirit in which the organization functions.

The experiments carried out at an accelerator laboratory are based on a simple principle: the accelerators provide beams of very energetic particles. At the CERN accelerator, these particles were protons. One of the aims of the scientists using accelerators is to determine the internal structure and other characteristics of the proton.

Say you want to find out the properties of some hard object. You may want to know how large it is, whether two such objects attract or repel each other, and what the inner structure of the object is. If it is big enough, you could measure the size directly and study the attraction or repulsion by putting two of them close together. However, because protons are too small to perform such direct tests, you use what is really a most primitive method. A fast-moving beam of protons is hurled against a target, an assembly of stationary objects of the same or similar kind—for example a sample of hydrogen, since the nuclei of the hydrogen atoms are protons. You observe what emerges in different directions from the target. Sometimes the incoming proton has collided with a proton in the target, like two billiard balls colliding, ejecting the target protons in certain directions. From these occurrences one can deduce the size of the protons and the forces between them. But other more interesting processes are also taking place. When two protons hit each other, new particles are created. Actually, this is also what happens when two ordinary objects hit each other at high speed—sparks may fly off. The light of these sparks is created by the collision. New entities such as mesons or pairs of particles and antiparticles are created when the collision energy is high enough.

As discussed earlier, the existence of mesons was discovered in the forties in cosmic-ray research. The name *meson* indicates that its mass is somewhere between the electron mass and the proton mass. Today many kinds of mesons are known, the most important of which are the so-called pi-mesons, or pions. All mesons are short-lived entities; the charged pi-mesons exist only for a few hundredth parts of a microsecond (a millionth of a second), some of them even for a much shorter time, before they decay into other particles. There are also uncharged pi-mesons, which exist for a mere fraction of a billionth of a microsecond and then decay into two light quanta. They play an important role in particle physics.

The existence of an antiparticle for each particle was first guessed at by Dirac in 1931 and then found in nature, as mentioned earlier. The most exciting feature is that a particle-antiparticle pair can be created by a sufficient concentration of energy. Furthermore, when a particle and an antiparticle meet, a little explosion takes place: they disappear and their energy appears as a light burst, as a burst of mesons, or in other forms of energy.

The collision of high-energy beams with matter produces such high concentrations of energy that large numbers of particle-antiparticle pairs are created. This is why antimatter plays an important role in high-energy physics. It is even feasible to separate the antiparticles from other collision products, so that one can produce a beam of antiparticles—for example, a beam of antiprotons. Detailed study of the scattered protons and of the newly created particles and antiparticles enables one to learn about the inner structure of the proton or the neutron.

Essentially, three types of experiments are aimed at finding out what emerges from the collision of a beam of particles with a piece of matter containing protons and neutrons: the photographic emulsion method, the track chamber method, and the electronic experiments method. The first exposes special photographic emulsions to a beam. When the protons hit atomic nuclei in the emulsion, they produce new particles that can be seen in the form of microscopic tracks in the emulsion. This method was perfected in the 1950s by Giuseppe Occhialini and Cecil Powell in Bristol, who used it to study the effects of cosmic rays.

The track chamber method exposes large vessels filled with an almost-boiling fluid—usually liquid hydrogen—to the beam. Because the nuclei of hydrogen atoms are protons, this method allows one to study the collisions of protons with protons. (Sometimes it is interesting to have heavier nuclei as targets; then propane, freon, or xenon is used as the liquid.) Hydrogen liquefies at very low temperatures, so such chambers need elaborate cooling devices. When a particle crosses the container or when another particle emerges from a collision of the beam with an atom in the liquid, small bubbles are formed along the path of the particle. These rows of bubbles are photographed. This is why such devices are called bubble chambers. In order to prevent bubbles forming all over the container, the liquid is kept under pressure. When the pressure is released, bubbles form where the liquid was disturbed by the particles. That release is made to happen just when the beam enters the chamber. A loud boom is heard in the laboratory every second or so when the pressure is released.

In electronic experiments the particles emerging from the target are registered with electronic devices called counters or spark chambers, where the particles produce electric discharges. Much ingenuity has gone into inventing electronic sensors to detect particles, follow their paths, and identify their properties.

Indeed, there was no lack of interesting problems to investigate when the CERN scientists started using the accelerators. But the beginning was difficult because in the early days, before I came to CERN, most efforts were concentrated on the construction of the large proton-synchrotron (PS). The proton beam at the PS had an energy of 27 GeV, an energy far above that of other existing accelerators at that time. (Within half a year, however, the Brookhaven laboratory finished the construction of a similar accelerator.) Too little effort had been spent preparing necessary instruments to detect the effects of collisions. Thus, the exploitation was slower than it would have been had the preparation of experiments started earlier.

When planning the activities of CERN, groups of physicists within and outside the organization proposed a large number of experiments to be performed at the two accelerators. The challenge was to select the most interesting experiments. Obviously, the physicists within CERN were eager to use the machines as soon as they were operable. But CERN was supposed to offer opportunities to any European group and even to guest groups from nonmember states if the proposed research showed promise of great interest. Since not all of the proposals could be executed, a process that would result in reasonable and fair decisions was desperately needed.

A Nuclear Physics Research Committee was proposed for that purpose. The term *nuclear* was not quite appropriate, since the main topic of research at CERN was subnuclear physics, the internal structure of the nuclear constituents. (But, as I have said, the word *nuclear* persists even in the name of the insitution, leading to the mistaken inference that CERN deals with nuclear power. Today the institution is called "CERN, the European Laboratory for Particle Physics" to avoid the confusion.)

Many discussions were held as to whether there should be subcommittees for the three kinds of experiments. The emulsion and track chamber methods lent themselves more easily to research outside of CERN because emulsions and bubble chamber pictures could be evaluated elsewhere. Electronic experiments, however, must be carried out and evaluated on the spot, except in cases where the results can be transmitted by computer lines to outsiders. Because we did not have such computer hookups at first, the evaluation of electronic experiments away from CERN was not in general use until much later.

Probably too much time was wasted in these discussions because the European physicists did not have enough experience running large research institutions and the acting director, John Adams, was not a physicist. By the time I took over the directorship, the situation had gotten rather confused. I realized that I was not yet acquainted with the subtle intricacies of the laboratory and the character of the influential persons. I knew even less about the technical details of the machines and detectors and next to nothing about the administrative and financial setup. I needed someone to help and advise me in all these matters, somebody with whom I could work intimately, whom I would respect and who respected me. He would have to be someone who complemented my abilities in theoretical science by his experience in the technical and organizational aspects.

I am aware that I am not a good psychologist. Often in my life, I have overlooked the negative qualities of a person and seen only the positive side. But my choice of Mervyn Hine as my closest collaborator was an unmitigated success. He had worked with John Adams on the construction of the PS and was a member of the directorate when I came to CERN. It was therefore easy to have him work with me. I could not have known at first how well I had chosen, but I am now keenly aware that without Hine my directorship would most probably have been a failure. I would have acted in a vacuum. Hine was always ready to point out to me both the difficulties and the opportunities—in short, the realities—of any situation. In addition, his abilities in financial areas and his experience in preparing budgets were important to me, since I have no aptitude for such matters.

The most important thing was our personal relationship, one based on complete trust and mutual understanding, perhaps because our abilities were unique to each of us. A strong sense of humor on both sides helped a lot. We were able to talk to each other without any inhibitions. Often, I asked him to come to my house in the evening to talk over some decisions about which I felt uneasy. Sometimes I would weep on his shoulder. One of his remarks in such cases was, "Put your regretter on zero." It was his motto. He felt that once a decision was made, there should be no looking back. In spite of our amiable friendship, however, we often had fierce arguments. Mervyn was a strong and somewhat abrasive person who did not mince words when he disagreed. I appreciated this, but some others found him difficult to take, and he was not universally liked. I am a person who tries to avoid open confrontations. The net effect was that most of the staff gave me the credit for the measures they approved of and blamed the others on Mervyn. Actually, we did everything together in full agreement.

Another important friend and advisor was Bernard Gregory. He was a French citizen, but he had an English mother and was fluent in both languages. I met him first during my stay in Paris in 1950 and got to know him well during his "compulsory year" in the United States, which he spent working with Bruno Rossi at MIT on cosmic-ray experiments. Since the early 1950s he had been an active member of the research team of Louis Leprince-Ringuet at the École Polytechnique, making important discoveries in particle physics by studying the effects of cosmic rays on matter. Intellectually and administratively, Gregory was the leading spirit in that team. Under his guidance, the Paris team constructed a hydrogen bubble chamber, which had just been brought to CERN when I started my director-ship. It was one of the most successful instruments there.

In 1963, I asked him to become CERN's director of research, a position in which, as a theorist, I needed an experienced experimental physicist. I first had Gilberto Bernardini and later Giampietro Puppi as research directors. When I left CERN at the end of 1965, Bernard Gregory became my successor as director general. Bernard was a wonderful character who died much too young in 1975. My rela-tionship with him was in some ways similiar to that with Mervyn Hine. We had full confidence in one another. He was easy to work with, and we achieved full agreement about the research program and the future plans for CERN. He had a talent for finding logical reasons for any decision, even when they were based only on intu-ition, as many of my own decisions had been. I referred to his way of arguing as Cartesian. He worked extremely hard on any task, never forgetting any details, in contrast to my more general, intuitive approach. (During a flight to a conference in Hamburg, I noticed him studying a document. It was the text and the historic background of Wagner's *Die Meistersinger,* which we were planning to see while in Hamburg. He was not leaving even that to chance.) When he became director general, he was able to realize all the plans that we worked out together.

Another valued co-worker at CERN was Charles Peyrou, a French physicist whom I first met when I was a guest professor at the Sorbonne in 1950 and he was working with Leprince-Ringuet. Charles impressed me as an excellent experimental physicist, but I was also attracted to his lively character and wit. I had arranged for him to come to MIT for a year to work with Rossi on cosmic rays at the same time as Bernard Gregory. At CERN he became the leader of the track chamber division, where a number of significant dis-coveries were made, and I had regular meetings with him to discuss outstanding problems.

Another of my colleagues was Theo Kröwerath, head of the trans-port service and thereby in charge of all sorts of transportation from cars to cranes. Now retired, Kröwerath, like many Berliners, was lively and active, with a quick intelligence and an excellent way of dealing with people. He had no scientific training whatsoever, but a very good understanding of the needs of the scientists and of the support staff.

After I had been at the laboratory for a while, I asked him to drive me from my home to the laboratory once a week so I could talk to him. Theo knew everything about what was going on with the mechanics, drivers, plumbers, cleaning staff, and others who formed CERN's vital support staff. He was an important source of unbiased information for me. In the somewhat isolated position of director general, one can easily lose touch with the people who maintain the efficient operation of the organization. Kröwerath could tell me about all the complaints and gripes I would never have heard in any other way. Thus, I was able to respond to problems directly and could help improve the morale among those who are more important for a successful laboratory than most people realize.

I also made biannual visits to the workshops where the instruments were assembled. I thoroughly enjoyed seeing how the complicated instruments were constructed, and I was happy that I could talk to many of the workers in their mother tongue. Whenever I heard that a special instrument was completed, I congratulated those who had done the job. I realized that such recognition coming from the director was important to people who often have to work long hours under great pressure and who rarely get the recognition their work deserves. In a similar spirit, I also tried to be present when a group of experimental physicists arrived at a critical moment in their work. My role model in all of this was Oppenheimer, who was never too busy to be an attentive colleague when he was director of the Los Alamos laboratory.

Once, just as I was about to visit one of the workshops, my secretary told me that a representative of the Italian government had arrived to talk to me. I let him know that I had no time just then because I was going to visit the workshops and that he would have to wait for two hours or come another time. I was not sorry when I heard that the secretary had told the staff about that episode. Today, more than twenty years after leaving the job, I still enjoy the personal loyalty of the CERN staff, and I am grateful for this gift. I still spend most summers at CERN, since I own a summer chalet in the mountains nearby. Whenever I ask someone from CERN to do some minor work on the house, they not only oblige me gladly but never send me a bill, in spite of my insistence.

Antonino Zichichi is another valued friend at CERN. A physicist who performed some interesting experiments while there, he discovered the antideuteron. The deuteron is the simplest composite nucleus consisting of a proton and a neutron bound together. The existence of the corresponding antiparticle shows that antinuclei and antiatoms do exist.

A Sicilian, Zichichi has a sparkling character, and he bubbles over with ideas of all sorts, not only about new experiments but about activities to promote science in Italy and all over the world. Although some found him autocratic, I was impressed by what he achieved.

One of Zichichi's most successful ventures was the founding of a summer school for particle physics in Erice, a romantic medieval town on the top of a rocky elevation in western Sicily. Summer schools are some of the most pleasant institutions in science. The first such school was founded in Ann Arbor, Michigan, in the 1920s, and after the Second World War many institutions all over the world took up the idea. Usually the programs were held in some pleasant or romantic location so the participants could enjoy the pleasures of the surroundings as well as the intellectual stimulation of the lectures and working groups.

The school in Erice has special attractions. Nino is able to select the best and most successful scientists as speakers. He is very careful in the selection of the students and guests who are privileged to attend. Nino worries about everything—the scientific programs, the discussion, and the regular attendance of the students. He even concerns himself with the quality of the food in the restaurants of the town. If he considers one of them below standard, he declares it out of bounds.

He is also a great master at getting money for his enterprises, often without much concern about the nature of the sources. Due to his efforts, the summer school is supported by many institutions, including the North Atlantic Treaty Organization (NATO), in spite of its being a military organization. I frequently participated as a speaker or a guest. The proceedings were always published in book form and were widely used as texts. In later years the Erice school expanded greatly. Today it holds working weeks not only in particle physics but in other fields of physics, biology, and medicine and also concerns itself with the problems of war, peace, and arms control.

Another important Italian physicist at CERN was Carlo Rubbia. Carlo was very young when he came to CERN, but he impressed me as extremely intelligent and full of ideas for new experiments. In spite of his youth, I decided to give him a rank high enough for him to realize some of his own plans. I had to convince the senior physicists that Rubbia's age was not a drawback because he was capable and full of good ideas. It is true that he was, at that time, perhaps a bit too full of ideas and was often not patient enough to finish one experiment before starting another. But his discovery of the carriers of weak interaction in 1983 brought him the Nobel Prize, the first one awarded for work done at CERN. He became director general in 1989.

I always found it exciting to deal with my Italian colleagues. Some people in the theoretical division complained that they could not work because of the loud conversations of Italian theorists in adjacent rooms. We decided to put soundproof walls between offices, but the absorbing material had a strange property: it seemed to absorb all languages except Italian.

My task at CERN was made easier because the idea of European cooperation had strong support at that time, and Europe was at the height of its "economic miracle." It was much more difficult for my successors to get the necessary financial support. The financial situation is no longer so splendid, and the idea of European cooperation is widely enough accepted today that CERN does not play an important symbolic role anymore.

I had another advantage: I was not a citizen of any member state, so I could not be accused of any national bias. My American experience combined with my European background was a great help, as was my reputation as a knowledgeable scientist. I was lucky also to discover a talent for dealing with large groups of people that I had never known I had. By watching other successful administrators,

I had learned that the trick was to run a place according to your own ideas but arrange it so that your co-workers thought the ideas had come from them. Of course, nothing was done without long discussions with Mervyn Hine, Bernard Gregory, Charles Peyrou, Theo Kröwerath, and others.

My main interest always remained the scientific program and the decision-making process affecting the various research projects. I was chairman of the Nuclear Physics Research Committee, which discussed the proposed projects and made the final decisions. The subcommittees for the different methods were helpful in clarifying the proposed experiments and fitting them into the operating schedules of the accelerators, but they played a minor role in decision making.

To emphasize the spirit of European cooperation, I insisted that anyone who entered CERN be regarded as a European and no longer as a citizen of some nation. My idea was to minimize national concerns in the scientific program. Unfortunately, it was not possible to eliminate them completely. All too frequently one nation complained that its physicists were discriminated against in the choice of experiments. Sometimes we were forced to let national interests override scientific judgments in order to avoid disturbing conflicts.

During the 1950s the Soviet Union also created an international high-energy physics laboratory, located 100 kilometers north of Moscow in Dubna. This venture was not as successful as CERN for several reasons, not the least of which was the domination of the Soviet Union. Another reason was that Dubna did not use the new strong focusing design for its accelerator, so it had less available energy. And basically, the accelerator suffered from the backward state of engineering in the Soviet Union at that time. The stifling results of Soviet domination over Dubna reconfirmed our belief that the United States should not be a member of CERN. American participation would have overwhelmed the European member states.

Since we had no members from the countries of the Communist bloc, CERN could not be considered truly European. For me this was a deplorable shortcoming, but there was no way to correct it during this period of cold war tensions. I still clung to the ideal of an all-European CERN, however, and tried to overcome the exclusion of the East European countries in other ways in spite of difficulties on both sides. There were excellent particle physicists in the Soviet Union, Poland, East Germany, and Czechoslovakia. But the West German government did not like the idea of having East Germans at CERN, and the Soviets were reluctant to allow their physicists to spend time in the West. After extremely difficult negotiations we finally managed to bring a number of Eastern scientists to CERN, including Soviet ones. A Polish group under the leadership of Marian Danysz succeeded in making a fundamental discovery at CERN. They proved the existence of the so-called double hypernuclei, that is, nuclei in which two of the nucleons are in special long-living excited states, which justifiably are called "strange states." Poland was subsequently given the honor of being an observer state with representatives on the CERN council, but not only as a reward for that discovery.

We also began more systematic negotiations with Soviet high-energy laboratories. Eventually, the exchange of scientists with the Dubna organization was increased, but our main aim was to collaborate with a large new laboratory in Serpukhov, 100 kilometers south of Moscow. There the Soviets had begun an ambitious project to build an accelerator with an energy of 70 GeV, almost three times the energy of the CERN machine. The Serpukhov accelerator profited from the improved engineering capabilities of the Soviet Union.

This accelerator remained the highest-energy machine in the world until CERN and the United States each built a 400-GeV accelerator in the early 1970s. CERN contributed to the exploitation of the Serpukhov accelerator not only to further international collaboration but also to ensure CERN scientists' right to explore a hitherto unexplored energy region. We agreed to deliver certain high-technology devices with which the Soviets had less experience, such as a system to eject the fast particle beam out of the accelerator where it could be much better exploited. We also provided a radio frequency separator that separates different secondary beams containing different particles, and French physicists constructed a special large bubble chamber to be used in Serpukhov. In exchange for this assistance West European physicists would be admitted to the Soviet facility to perform experiments at energies unavailable anywhere else at that time. I initiated these exchanges while I was director, but they were formally agreed on and executed by my successor, Bernard Gregory, from 1965 to 1970. I was gratified to have laid the groundwork for opening and widening East–West collaboration in high-energy physics. The East European and Soviet physics communities clearly appreciated it, and I enjoyed the many close personal relationships that grew from these activities.

By 1964 the Soviet Union was considered a partner in high-energy physics. As a symbol of this relationship an international Rochester-type conference was held in Dubna in the summer of that year. During a boat trip on the Volga, Andronik Petrosyants, the chief of the USSR State Committee for Atomic Energy, invited me to have a drink with him. He filled two enormous glasses with strong vodka and proposed a toast to the collaboration. He drank the whole glass down in one gulp, but I could only drink a small part of mine. He said, "Oh, I see that you only want a little collaboration?" What choice did I have? In the interest of international cooperation I had to drink the whole glass. I admit that I had a hard time walking down the gangplank when the boat landed.

As noted at the beginning of this chapter, I arrived at CERN a comparative novice in particle physics. I thought it was part of my job to improve my knowledge in that field, so I forced myself to devote some time to my education by announcing a series of weekly lectures entitled "An Introduction to Particle Physics." I had discovered long ago that the best way to get a basic understanding of anything was to teach it. I also wanted to demonstrate to the staff that their new director was not interested only in the problems of administration, that physics was also important to him. The lectures were well attended. I kept them on a level that the less knowledgeable scientists and engineers could follow, and I tried to make them exciting so that those who were familiar with the subject would find it interesting to see how I presented the problems. I believe that these lectures contributed something to the spirit of CERN as it grew from a construction site into a functioning laboratory.

It is not hard to attract an audience to a lecture on particle physics. The field exerts a special attraction because it penetrates into the innermost structure of matter. The late 1950s and early 1960s were extremely fertile periods for particle physics. Enrico Fermi and his collaborators were the first to discover, in 1952, that the proton and the neutron are not "elementary" particles. They found that when enough energy is transferred to them, these entities change their quantum states. They become excited to higher quantum states (meaning higher energy) just like atoms or atomic nuclei, only on a much higher energy scale. After Fermi's discovery, other laboratories, at Berkeley, Cornell, and elsewhere, found more excited states of the nucleons. The energy differences between atomic quantum states are only a few electron volts or less. The quantum states of nucleons differ by much larger amounts of energy, on the order of hundreds of thousands of electron volts.

As mentioned in chapter 2, the ensemble of excited quantum states of a system is called its "spectrum." For example, the catalog of excited states of the hydrogen atom is called the hydrogen spectrum, and every atom similarly has its characteristic spectrum. When the excited states of atomic nuclei were found in the thirties, a second "spectroscopy" was discovered: the spectra of atomic nuclei. Now, initiated by Fermi's discovery, there arose a third spectroscopy: the spectrum of the nucleon (the proton and the neutron).

The existence of excited states always points toward an inner structure. The quantum states of atoms were recognized as different ways in which the electrons assembled around the nuclei; the excited states of nuclei indicated the different ways in which the nucleons join to form nuclei. Thus, the excited states of the nucleons suggested that they are also made up of certain constituents.

I discussed earlier how when an atom in an excited state returns to its normal (ground) state, it gets rid of the surplus energy in the form of light, in a process called the emission of light by atoms. To give an example in practical terms: the current in an electric light bulb excites the atoms into higher states, and their return to ground state produces the light. A nucleus in an excited state also returns to its ground state by getting rid of its surplus energy, usually in the form of light of rather high energy, the so-called gamma rays; but sometimes it accomplishes this by a special process called radioactivity. In this case the nucleus emits a pair of particles, an electron and a neutrino. When this happens, the atomic nucleus changes its charge. The nuclear charge determines the number of electrons and therefore the character of the atom. In this way a radioactive process changes one type of atom into another. The existence of a neutrino was suggested by Pauli as early as 1930 in order to explain some surprising features of radioactivity. In later years more evidence accumulated as to the existence of that elusive particle.

Excited nucleons also return to their ground state by the emission of light, but primarily by the emission of mesons. Of course, the amounts of energy of excited nucleons are immensely greater than those of excited atoms or nuclei. So are the energies of the emitted light or mesons. It turned out that the mesons themselves also appear in different quantum states; a meson spectrum exists.

In the 1960s the properties of the excited states of the nucleon and of the mesons led two theorists—my old student Murray Gell-Mann and, independently, George Zweig, who was at CERN at that time—to hypothesize that protons and neutrons are made up of three new sorts of elementary particles. (Gell-Mann called them by the ugly but memorable name "quarks.") Indeed, most of the properties of the observed excited states of the nucleus could be explained by that hypothesis. Gell-Mann's triumph was the prediction of the existence and the energy of a hitherto unknown special state of the nucleon— the so-called $\Omega$ particle—which was indeed discovered soon afterward at the Brookhaven National Laboratory. He received the Nobel Prize for this work in 1969.

Furthermore, the different quantum states of the mesons—the meson spectrum—could be interpreted as different combinations of a quark and an antiquark. This explained many of the meson properties—for example, their short life span. The quark and the antiquark that form the meson are bound to annihilate each other. The products of this annihilation are electron–neutrino pairs or light. Indeed, the mesons are found at the end of their life span to "decay" into these particles. The theory also explained the emission of mesons by an excited nucleon. The energy contained in the excited states is high enough to create quark-antiquark pairs, appearing as mesons.

The emission of electron–neutrino pairs produced when excited nucleons return to their ground states is not the only manifestation of radioactivity. Some mesons decay by emitting such pairs. When excited protons, or mesons, emit electron-neutrino pairs, a new feature appears. Enough energy is available that, apart from a pair of ordinary electrons and a neutrino, another pair of particles is emitted: a heavy electron and a neutrino. The heavy electron—the muon— is 200 times more massive than the ordinary one. The muon is an unstable particle. It decays by a radioactive process into lighter particles. Thus, the processes of radioactivity become more and more multifaceted.

The high concentrations of energy produced in the collision of high-energy beams with matter create particle-antiparticle pairs. This is why antimatter plays an important role in high-energy physics. It is even feasible to separate certain particles and antiparticles from other collision products so that one can produce a beam of antiparticles, for example, a beam of antiprotons—a most remarkable achievement. The detailed study of the scattered protons and of the newly created particles and antiparticles enables one to learn about the inner structure of the proton or the neutron.

When research began at CERN after the completion of the accelerators, it was clear that the first experiments should focus on how the energetic protons were scattered when they hit hydrogen or other materials and what kind of particles emerged from these collisions. As expected, at the higher energy level available at CERN, more mesons and antiprotons were produced. The emerging particles could be collected to form secondary beams of mesons or antiprotons. These secondary beams were used in a way similar to the primary beam in order to discover what happened when mesons or antiprotons hit a target.

Once a new energy region is opened up, new effects are bound to result. Some of these were found in the first experiments at CERN. A typical example was a change in the character of the impact of one proton on the other when the energy of the impact increases. The proton then appears "mushier"—that is, the interaction spreads out over a larger volume, without becoming altogether stronger. One finds fewer protons scattered at large angles. This effect is called "shrinking of the diffraction peak."

Another important finding resulted from comparing the scattering of fast protons and fast antiprotons by matter. It was discovered that with increasing energy, the scattering becomes very similar, whereas at lower energy, the particles and the antiparticles scatter very differently. The theorists had predicted this result. The consequence of general conclusions, it is called the "Pomerantchuk theorem."

Perhaps the most important discoveries of CERN in the sixties were made with bubble chambers. The first effective chamber at CERN was built by the French under Bernard Gregory's guidance. Filled with liquid hydrogen, the chamber was eighty centimeters long. (Today we would consider this a very small chamber. The longest one, used twenty years later, was four and a half meters long.) Still, the French chamber was an excellent research instrument. When the chamber was exposed to the energetic protons beam of the accelerator, the researchers could study their collisions with the stationary protons in liquid hydrogen. About six million pictures of proton–proton collisions and meson–proton collisions were taken with that chamber. Special scanning apparatus had to be invented and constructed to facilitate the evaluation of these picures and to reconstruct the details of the collisions. As it happened, these machines were operated mostly by female workers, referred to in those days as "scanning girls," which was by no means a pejorative term for these highly skilled workers. Physicists spent much time observing the details of the scanning process. Not surprisingly, many marriages occurred between the people working with the scanning machines. Apparently the scientists were not only interested in the properties of the particles in the chamber but also in the particulars of their collaborators at the scanning desk.

When analyzed, the pictures revealed the existence of many more excited quantum states of protons and mesons than were known before. The bubble chambers at CERN greatly contributed to the "third spectroscopy," that is, to the knowledge of the excited states, their special properties, and the particles emitted when they lose their energy. A number of interesting results emerged from these studies confirming the quark structure of protons and mesons.

Another area of intensive study involved the neutrino. Because it is without charge, without discernible mass, and subject only to weak interactions, the neutrino is able to penetrate large amounts of dense matter. Neutrinos produced by radioactive processes in the interior of the sun penetrate the sun and can be observed on earth even at night, when the sun is on the other side of the planet. Since they easily penetrate the bulk of the earth, the only way to observe neutrinos is to catch one of the rare events when the neutrino interacts weakly with an atom in a large massive detector.

At CERN we tried to answer an interesting question: Is the neutrino that is emitted together with the heavy electrons the same neutrino emitted with the ordinary electrons, or are there several types of neutrinos? Of course, the American physicists using the similar accelerator at Brookhaven were also deeply interested in that question. The Americans beat CERN in that competition, even though their accelerator was finished six months after the one at CERN. A group headed by Leon Lederman, Melvin Schwartz, and Jack Steinberger could show in 1962 that the muon neutrino is indeed different from the electron neutrino. In 1988 they received the Nobel Prize for this work.

At the CERN accelerator it was more difficult to get a neutrino beam, a fact that was recognized too late because of incorrect calculations. A year later, however, CERN was able to accomplish two important technical innovations. One was the extraction of the primary proton beam from the accelerator so that the collisions of that beam with matter took place outside of the accelerator, providing much more intense secondary beams in particular beams of pi-mesons that decay into the neutrinos under investigation. The other invention was Simon van der Meer's so-called neutrino horn, a device that concentrates and focuses the pi-meson beam before it produces neutrinos.

CERN confirmed Brookhaven's findings and obtained 200 times more neutrino events, enabling the CERN researchers to get much valuable quantitative information about the two neutrino types.

Much was accomplished during my tenure at CERN, but more could have been done. Because the European scientists were acquiring experience so slowly, we were prevented from fully exploiting the possibilities offered by the new equipment. This led to some disappointments. An awkward moment occurred in 1963 when a few CERN physicists thought that they had discovered the carrier of the weak interaction. They reported this sensational result to an international conference, and it even made the *New York Times*. But it was wrong—they had misinterpreted their findings. The energy available at that time would have been much too low to produce the particle they thought they had discovered. Twenty years later the carrier particles of the weak interaction were actually found—at CERN!

It took a decade before CERN reached a level comparable, and in some respects even superior, to the American laboratories. Also, after the European scientists had acquired the necessary experience of working with big machines, a definite difference, albeit a surprising one, remained between the European and American approaches. The European method of constructing accelerators generally was less risky and safer, even though initially more expensive. By contrast, the Americans seemed imbued with a pioneering spirit that included much more improvisation and relied on subsequent changes if things did not work out. In general, the Europeans chose the engineering approach, the Americans the approach of experimental physicists. One would have predicted it to be the other way around—the Americans being traditionally more engineering minded and the Europeans more individualistic. Good engineers may have been easier to find in Europe, because many fewer of them worked for the military there. Whatever the reason, the result was that the European accelerators and their instrumentation were more carefully designed, and they worked more reliably and suffered fewer breakdowns.

A number of outstanding physicist-engineers working at CERN and at other European high-energy physics laboratories were extremely ingenious and productive in inventing new ways of detecting particles and improving accelerators. They and their colleagues established a tradition of machine construction frequently superior to the efforts in the United States. George Charpak is one of those talented people. I had the splendid idea of asking him to join the CERN staff in 1962. He is the most ingenious designer of instruments I have ever met, and CERN owes him a lot. His inventions of specialized apparatus designed to detect and identify particles often gave CERN considerable advantages over other laboratories. Over the years he developed many ways of refining detectors.

One of the important requirements of particle physics is the continual invention of better means of detecting weak radiations of all sorts. This is absolutely necessary because one often needs to observe very weak effects that are of great significance. Working at the limits of observability under most trying and unusual conditions, particle physicists are forced to develop precision instruments. Many of these new, sensitive detectors are also important in other fields of science such as biology, medicine, geology, environmental science, and materials science. In this way, high-energy physics contributes decisively to other sciences, a fact that is often pejoratively called "spin-off." Actually it is one of the most important contributions of particle physics to other sciences and to technology.

Today Charpak's devices are in use in laboratories all over the world. When I hired him, I could not have foreseen how productive he would be. I think appointing him may have been one of the best things I did for CERN. I have always enjoyed Charpak's wonderfully open mind and his interest in all aspects of human relations and all kinds of people. I still meet him often when I return to CERN, and it is always a pleasure to discuss politics with him and hear his ideas about how to improve the sorry state of the world. We agree on many points, but he has never failed to introduce me to new perspectives based on his broad personal contacts with the world at large.

Simon van der Meere, whom I mentioned before, was another most creative CERN physicist-engineer. He invented the neutrino horn as well as many other highly ingenious instrumentations. I dubbed him the "Maxwell demon of the twentieth century" because of the following invention. He was able to reduce considerably the spread in energy of the protons in the beam circulating within the synchrotron. He did this with a contraption that detected protons that were moving too fast (or too slowly) on one side of the circular path. When such conditions existed, his device would send an electronic message to the other side either to slow the protons down or to accelerate them. In the nineteenth century James Maxwell explained why a mixture of a hot and a cold gas always results in a gas of medium temperature and never in a part of the mixture (say, the right side) becoming hotter, and another part (such as the left side) becoming cooler. He said that such a state of affairs would only be possible if a "demon" would send the fast molecules of the gas to the right and the slow ones to the left. Of course, there are no such "Maxwell demons," but van der Meer's contraption essentially performed just such a task. Van der Meer's "stochastic cooling," as it was called, contributed much to the discovery of the carriers of weak interactions. He therefore shared the Nobel Prize with Carlo Rubbia in 1988 for that achievement.

The full potential of the CERN laboratory was realized only many years after I left the directorship at the end of 1965. It has become one of the best—if not the best—laboratories in high-energy physics in the world. Two important decisions made during my tenure did have an impact on the subsequent development of CERN. One was a reform of the budgeting program. In most government-supported institutions, the budget is determined each year for the following year. This short-range process creates difficulties in large laboratories, where many projects can take much more than a year to be realized and the planning requires even longer.

The reform program, named the "Bannier plan," for the Dutch chairman of the working party that developed it, provided a four-year rolling budget period. This meant that each year the budgets for the following two years were approved and fixed. The budgets for the two successive years were provisionally determined and subject to changes in the following years. This flexible budgeting system made it much easier to plan ahead. Many American laboratories envied CERN for the Bannier system, which, despite its name, was largely conceived and put into practice by Mervyn Hine. Indeed, the Bannier plan was a major diplomatic triumph. It thwarted a tendency to keep the yearly budget constant, apart from accounting for inflation, a tendency the British government wanted to support. Actually, the budget rose by about a factor of two in the 1960s, almost 13 percent per year over and above inflation. Another advantage that CERN has enjoyed over other international institutions is that the budget has always been determined in Swiss francs and not in dollars. The devaluation of the dollar in recent years has had no negative effect on CERN's budget.

The second decision was even more important because it dealt with the actual research to be carried out in the future at CERN. It had become clear by 1962 that the two accelerators we were using would soon become obsolete and that plans for new larger and more efficient equipment had to be developed before that happened. Many formal and informal discussions among the staff and among European physicists led, in 1963, to the establishment of the European Committee for Future Accelerators (ECFA) under the chairmanship of Edoardo Amaldi.

Two main proposals were made for long-range planning. One was to build an accelerator with about ten times the energy of the one in use. This was technically possible at that time and was also being considered by the Americans. It certainly promised to open up new frontiers in particle physics, because there were many questions for which the answer could only be found at higher energy.

The other proposal was for an entirely new type of machine, a so-called collider, in which two particle beams would be made to collide head on. Such an arrangement was proposed as early as 1943 by the Norwegian Rolf Wideroe. The advantage is not hard to understand. When a proton is accelerated, it acquires more kinetic energy and therefore becomes heavier, as explained by Einstein's equivalence of energy and mass. This can have a huge effect. For example, the protons in the PS of that period at CERN weighed about twenty-five times more when accelerated than when they were at rest. This greatly reduced the effect of collisions between the accelerated protons and those at rest. When a heavy object collides with a light one, it transfers only a small part of its energy to the light object. The energy of the collision is much less than the energy carried by the heavy object. An exaggerated example may make this clear: Imagine that a moving car hits a fly. That collision transfers only a small part of the car's energy of motion to the fly. When a fast proton whose mass is twenty-five times greater in motion collides with a proton at rest, only about a fourth of the fast proton's energy goes into the collision. Three-fourths remains with the fast proton, just as most of the energy remains with the car when it hits the fly. But if two high-energy proton beams in which the protons have twenty-five times their rest mass collide head on, their full energy comes to play in the collision, just as it would if two cars were to collide. In order to reach the same collision energy with an ordinary accelerator hitting targets at rest, one would have to accelerate protons to about sixty times the energy of the big machine in use at that time.

In the CERN collider, the collision of two beams was to be achieved in the following way. A beam of protons is extracted from the proton accelerator and divided into two separate beams. Each beam is then collected and stored in two storage rings, or large circular pipes contained in magnets that keep them in circular motion, one beam going in the opposite direction of the other. The storage rings intersect each other at eight spots where the two beams collide. This arrangement was called ISR, for intersecting storage rings. There was one drawback, however: proton beams are very dilute. When two beams collide, many fewer collisions occur than when a beam hits a piece of matter at rest or penetrates the liquid in a bubble chamber.

A number of senior CERN physicists opposed the collider project. They argued that while an ordinary accelerator with ten times the energy might not produce collisions of such high energy, it would be more efficient. The high number of collisions achieved would allow the exploitation of secondary beams emerging from the collisions, whereas the secondary beams of a collider would be too weak for use. They feared that the construction of a collider would postpone or even prevent the funding of a more energetic accelerator, which admittedly was a necessity for the future of CERN. Others argued against the collider because the Americans at Brookhaven had decided against building one. To this argument I could only say, "Why did we build CERN if we are only going to do what the Americans do?" I was convinced that CERN should use its superior engineering abilities to pioneer the construction of a proton collider. It would be a new type of machine; so far only electron beam colliders had been built.

Hine, Gregory, and I, as well as some influential physicists outside of CERN such as Francis Perrin in France and Edoardo Amaldi in Italy, strongly supported the collider. We thought that, in spite of their low number, the much higher energy of the collisions would give us a head start, a window into the future until regular accelerators of corresponding energy would be available. I finally convinced the council to approve the collider's construction. Kjell Johnson, the designer of the machine, was put in charge of the construction project, which took place after I left CERN. When it was completed, the extraordinary capabilities of the CERN engineers were once more apparent. The machine worked even better than we had anticipated; the number of collisions was over ten times larger than expected.

Many new results were found at these high-collision energies. With the ISR, CERN scientists could have detected an important new particle called the J/Ψ-meson, which is composed of a special type of quarks. Unfortunately, it was overlooked because of the European team's lack of experience and later found by Burt Richter at Stanford and Sam Ting at Brookhaven. They earned the Nobel Prize for their discovery in 1976.

It was more than a decade before European physicists learned how to work with large accelerators and could be considered truly on a par with their American colleagues. The discovery in 1983 of the existence of the carriers of weak interactions—one of the most important discoveries made at CERN—was only possible because of the transformation of the larger proton accelerator into a collider of protons and antiprotons. The experience gained with the ISR enabled the CERN engineers and physicists to perform this feat in a relatively short time.

The CERN initiative to build the first proton collider paid off handsomely not only for CERN but for high-energy physics all over the world. With the ISR, CERN introduced a new way of producing high-energy collisions. Because of the expected low collision rate, we first assumed that it would only give us a glimpse into what happens at very high energies. But as a higher number of collisions were produced and more sensitive detectors were developed, it proved to be an extraordinarily valuable research tool. Today almost every new accelerator project is planned as a collider and ordinary accelerators are converted to colliders. What we once regarded as a window into the future has become the way of the present. I consider the strong support of the ISR my most important contribution to high-energy physics as director of CERN.

During the last years of my directorship, it was clear to us that the ISR would only be part of planning for the future of CERN. An ordinary accelerator with ten times the energy of the PS was an obvious necessity, together with a general program to improve the equipment. At my last meeting in December 1965 the council supported such an improvement program, the construction of the ISR, and a supplementary plan to prepare for the construction of a new giant accelerator called the super–proton-synchrotron, the SPS. The funds necessary for these projects set a record for CERN totaling nearly a billion Swiss francs over the following four years. This equaled, at that time, about a quarter of a billion dollars.

A significant step toward further internationalization of CERN came as part of planning for its future. It became clear that the ISR and SPS would need more space than was available on the CERN site. A strip of land across the Swiss border on the French side looked just right for an enlargement of our site, as had been foreseen by the CERN founders when the site was chosen. We explored the possibility of acquiring that terrain and were pleased to learn that France was ready to give it to CERN. The formal notification came in a document from the French government. A few days later, on 13 September 1965, I signed the contract with France establishing a truly international laboratory extending over a frontier. For me, this was a great moment in the history of CERN. We had hoped that General Charles de Gaulle would ride into the French part of the laboratory on a white horse at the opening ceremony, but unfortunately he was unable to attend.

At CERN, I had felt again the special bond that is so often forged when physicists work together. The circle included many people with whom I had professional relationships that ripened into close friendship. I learned a lot from Leon Van Hove, a Belgian theorist, and found Maurice Jacob to be one of the few theoretical physicists interested in establishing close contact with experimenters. As leader of the theory division, he always received me with grace and enthusiasm when I returned to CERN after relinquishing my directorship.

Probably the most penetrating mind among CERN theorists is John Bell, who is well known for his insistent striving for a better understanding of the foundations of quantum mechanics. In my discussions with him I often had to admit that my ideas about some fundamental principles of quantum mechanics were not sufficiently thought out. He admonished me for my fuzzy thinking with such charm and clarity that I not only always learned something but also thoroughly enjoyed the lesson.

I had known Tsung-Dao Lee, from Columbia University, before I came to CERN. I still meet him at CERN almost every summer and have always loved hearing about his new ideas on many of the yet-unsolved problems of physics. Furthermore, I have been impressed by his efforts to establish bonds between Chinese and American physicists and students.

Finally, I must mention Yoshio Yamaguchi, and his wife, who spent much time at CERN and became close personal friends of mine. Yamaguchi contributed much to the scientific atmosphere in the theoretical division at CERN. Later he became an international statesman of science when he was appointed chairman of the International Council for Future Accelerators.

One other encounter deserves special mention. In 1962, at the height of the Cuban missile crisis, while the world was waiting to see if the United States would be able to prevent a tragic confrontation over the Soviet weapons in Cuba, there was a knock on the door to my office. When I opened it, there stood Leo Szilard. "World War III is about to break out," he said, without preliminaries. "I am the first refugee from America. I have come here with my wife and have brought all my belongings. Now I need a desk and a little secretarial help."

Although I was a bit taken aback, I invited him into the office. We sat down and talked for a while, and I assured him that I could give him what he had asked for. But I suspected that he had something specific in mind, some project that he wanted to work on. It turned out to be nothing less than the prevention of a third world war. He said he wanted to talk to Nikita Khrushchev immediately. In the following days he launched a tremendous amount of activity, telephoning around the world and occupying all of my secretaries most of the time. As one of them remarked, he acted like a virus taking over the metabolism of a cell. A driver waited in front of his hotel to bring him to my office and take him back to his hotel when he wanted to leave.

After several days, as we all know, Kennedy and Khrushchev worked the problem out, and the Cuban missile crisis was peacefully resolved. Szilard's efforts were no longer necessary. However, a Leo Szilard cannot be held down. Wherever he is, he has to initiate something momentous. When he saw that war had been averted, he came up with a new proposal. If the European nations had a common laboratory of high-energy physics, there should also be a European institute of molecular biology organized along the same lines of CERN. I thought it was a great idea.

Szilard got right to work. Sitting in my office and using my special staff, he organized a meeting of the leading European molecular biologists. A few weeks later at their meeting—in my CERN office—he was able to arouse their interest and enthusiasm. Sure enough, after only a few years the European Molecular Biology Organization (EMBO) was established under the directorship of John Kendrew, a well-known British biologist. EMBO's headquarters are in Heidelberg, and the organization has become an important center of biological research.

Having proved that his stay in Geneva was not in vain, Szilard returned to the United States. My secretaries were relieved, but I was once again impressed by Szilard's extraordinary creative abilities, which had repeatedly led him to conceive of ideas and nurture them to fruition. Because EMBO was initiated at CERN during my directorship, I am often wrongly given credit for its existence, and I am glad for this chance to set the record straight.

After so many exciting developments the question must be asked: Why did I leave the job in 1966 just as CERN was beginning to flourish? It would probably have been good for CERN had I stayed a total of seven or eight years, especially since my five-year term as director general set a precedent and became the normal tenure for a director. I now think that five years is too short a period to accomplish everything one wants. It takes two years to get acquainted with the details of the job, and during the last year many people see the director as a lame duck. I left CERN for several reasons. First, I thought it was time to leave. I was satisfied that I had helped put CERN on a firm footing and was confident that my successor, Bernard Gregory, with whom I had such a close relationship, would continue the spirit I had tried to promote and would be able to realize our plans for CERN.

Perhaps a more important factor, however, was that Ellen was not too happy in Geneva. In Europe the position of the director general is so exalted that it prevents close friendships between his family and the rest of the staff. This is not so in the much less formal atmosphere of the United States. Also, the society in Geneva is rather exclusive and does not favor intimate contacts with the international set. Although we had made some friends—notably Martin Kaplan and his family and Victoria Winnecka, who were both working at the World Health Organization—Ellen was lonely and isolated. Furthermore, we missed our children and wanted to be with them.

Finally, there was the question of my own professional future. If I stayed on as director of CERN, my ties with the United States would have slowly eroded, and returning after my retirement would have been difficult. Spending my old age in Geneva was not an attractive idea to me. Going back to the United States gave me the opportunity to use my CERN experience in another environment and promised many years of productive activity in a different surrounding. For these reasons, we returned to Cambridge and MIT in 1966.

I did not lose my ties to CERN, however. For many years after leaving, I was a member of the Scientific Policy Committee, an international body consisting of physicists experienced in the field. Its purpose is to advise the council and the directorate about research and future planning. Also, Ellen and I spent almost every summer in Geneva to escape the humid heat of New England. As a consultant to the division for theoretical physics, I kept in close touch with the activities and further developments of CERN. Futhermore, I gave a course to the summer students who wanted an introduction to particle physics. The many staff members and guest physicists who also attended these lectures helped me keep abreast of what was going on in high-energy physics, at least as much as they informed the audience.

We knew when we decided to return to Cambridge that we would be making frequent trips to CERN. Rather than live in hotels or rented apartments during our visits, we sought a place of our own from which we could easily travel throughout Europe. During the last months of our stay we looked for a suitable piece of land near

Geneva where we could build a summer house for the subsequent years. We were very lucky in finding a charming place in the village of Vesancy in France, only ten kilometers from Geneva and six kilometers from CERN. Vesancy is an old village with a population of 300 people and over 1000 cows, with a twelfth-century castle that also serves as the mayor's office and school. In contrast to the other communities in that region, Vesancy has managed to keep the old-fashioned, rural character of an agricultural village. Although it is within commuting range of Geneva, Vesancy has resisted constructing any of those ugly concrete apartment buildings that house so many suburban Genevoises.

At first we had had a lot to learn about how things were done in Vesancy. One example was the purchase of our property. It did not take us long to find the site we wanted. We fell in love with it immediately and went to a real estate agency of the neighborhood to make arrangements to buy it. We were told that there was nothing for sale there. Disappointed, we asked William Gamble for help. As an administrator at CERN, he had become an expert in the customs of the region. He advised me to visit the mayor with Ellen, introducing myself as director of CERN with the intention of talking about the problems of the region. He explained that it might be prudent to praise the fifteen CERN employees who came from Vesancy, but under no circumstances should there be any mention of our hope of buying land. Then, he said, after several weeks we should return and slowly begin to talk about how wonderful we thought his village was. Then, when we sensed the moment was right, we could say that, in fact, we had found a *petit paradis* we would like to buy so that we could build a little house.

We followed this plan of action to the letter. After a proper interval we finally brought up the subject of the little house. It went just as our friend had predicted. Over a glass of the local red wine the mayor said, "Mais oui, nous sommes honorés"—in other words, no problem. He would arrange everything. Indeed, a few days later in his office he introduced us to the owner of the land, who was ready to sell for a reasonable price.

Our land is on the slope of the Jura Mountains, overlooking the village with its church and castle and the reddish roofs of the old houses. We have a marvelous view of the Alpine chain, and when no clouds hide it, we can see the white pyramid of Mont Blanc sixty kilometers away. The frequent rains result in lush, green meadows and healthy pine forests that surround our little chalet. We bought six acres of land, including all the meadows visible from the house to protect our view from construction. Taking care of the grass is easy because one of our neighbors is a cattle farmer. His herd grazes in our meadows, and he cuts the grass to make hay for winter feed. It is the custom to charge for the use of the meadows, but we do not ask for compensation. We value his friendship much more than the money he might pay us.

Another close friend and neighbor, René Ducret, works at CERN. He is now a master plumber, but he began as a member of the night cleaning crew. He ascribed his advancement to my interference. While still a cleaner, he asked me to use my influence to improve his working conditions. Actually, I did not do much to help him, since I did not consider it appropriate for a director to favor personal acquaintances. Instead, his efficient work and his eagerness to learn induced his boss to send him to one of CERN's training courses for specialized work. But he continues to believe that my influence made the difference. Over time we have become good friends, and he efficiently and lovingly cares for our chalet while we are in residence as well as in our absence.

We have become friendly with a number of people in the village. Often as I am working on my terrace, I am interrupted by some of my village friends who have stopped by to *bavarder* (gossip) and lift a few glasses of the local red wine. These friendships are easy and warm, and there is no feeling of class differences. The people of Vesancy speak a French almost as rich as that of city people. In Austria, and in many other parts of Europe, a peasant from a village would use a much more primitive language than an educated city dweller.

I have enjoyed these friendships, and I was greatly pleased when, in 1972, Vesancy gave me an honor that I value at least as much as any of my scientific awards. I was made *sappeur-pompier honoraire* (honorary fireman) in a big celebration in the old castle. During the ceremonial part of the evening I was given a fireman's helmet and a diploma, and a little girl in a white dress gave Ellen a large bouquet. There were lots of flowery speeches, and a great deal of wine was consumed.

Over the years we continued to enjoy our property immensely. Every time we returned, we found it even more charming than we had remembered. For me it is almost as beautiful as Altaussee. Wonderful walks through the woods and hills begin directly at our door. No new construction is visible from our place; everything is as it was when we bought it, except that the castle has received a new roof. Life is quiet and relaxing there, and the people are friendly. It is France at its best.

# TWELVE

⊠

# Back to America

RETURNING TO CAMBRIDGE was not as hard as we had expected. We reclaimed our house, which we had rented out in our absence. At Christmas in 1965, when our children and some of our friends assembled to celebrate the holidays, we already felt very much at home again. It was easy to take up our life in Cambridge. Soon we were going to parties, concerts, and theater performances in town, and once more there was chamber music at home. Although we missed the Swiss and French mountains, which we loved, the New England landscape had its own peculiar charm. We spent our weekends in the country—in the winter on skis, in the spring and fall on foot. We noticed that one thing had changed in our absence. In the late thirties, when we had first started to explore the American countryside, we had met very few people on the trails, and most of those had German accents. Now we found that Americans had discovered the joys of the outdoors, and many more people were on the trails, mostly from the younger generation.

At MIT, I was happy to find that I could devote more time to teaching, research, and the study of the new developments occurring at that time. The decade between 1965 and 1975 was an unusually fertile time in physics, with many exciting new insights and discoveries. There were several that particularly captured my interest and enthusiasm. First was the steadily increasing evidence for the quark structures of protons, neutrons, and mesons. All doubts about the existence of quarks disappeared when two MIT physicists, Jerome Friedman and Henry Kendall, jointly with Richard Taylor from Stan-

254

ford, were able in 1968 to "see" the quarks within protons and neutrons.

When we use our eyes to see, a beam of light impinges on an object and is scattered by it. The scattered light enters our eyes, where nerves in the retina send signals to the brain. Friedman, Kendall, and Taylor "saw" the quarks by means of a highly energetic beam of electrons produced by the Stanford linear accelerator. The beam was directed at hydrogen, containing protons. The quarks inside the protons scattered the electrons, which were then detected by electronic eyes. Just as our eyes determine the shape, color, and other properties of objects from the light scattered by them, these artificial eyes established the existence and properties of the quarks by analyzing the scattered electrons. Many such scattering experiments were performed afterward in most high-energy laboratories. Besides electrons, beams of muons and neutrinos were also used to be scattered by quarks inside the proton or neutron. These experiments were called "deep inelastic scattering." It was thus established that quarks are real particles and not just inventions of our mind.

Another captivating insight gained at that time was the understanding, at least in principle, of the forces that keep the quarks together within the protons, neutrons, and mesons, three types of particles often jointly referred to as "hadrons." This was based on a generalization of quantum electrodynamics, the theory of electromagnetic forces between charged particles. A sketch of this theory is necessary for an understanding of the new insights. The electric charges are the sources of the electromagnetic field, and the field acts on the charges. This idea accurately describes the attraction between positive and negative electricity. One charge produces a field, which then acts on the other charge, forcing it to move toward the first charge, if the two charges are opposite, and away from the first charge if the two charges are equal. Moreover, a vibrating charge produces a propagating, oscillating field, a light wave, which if it arrives at a wire—the antenna—sets up vibrations of the electrons within the wire. If it arrives at the retina of our eyes, it sets the electrons in motion, an effect that is transmitted by nerves to the brain: we see the light.

In the new theory it is not the electric charge but a "definite attribute" of the quarks that creates fields, which are not electro-magnetic. They are called "gluon fields," because they represent the glue that keeps the quarks bound within the hadrons. The attribute that produces gluon fields is called the "color" of the quark, although it has nothing to do with real color. This property comes in three varieties. To use such a threefold attribute as a source of a field required an ingenious generalization of the quantum theory of fields that is appropriately called "quantum chromodynamics." In practice, this theory proved difficult to work with, much more so than quantum electrodynamics. But the results derived from it described astonishingly well the properties of hadrons and their excited quantum states. Many theoretical physicists all over the world contributed to the development of quantum chromodynamics in the early seventies. It is hard to assign the credit to one person, but the advances started with a seminal paper by Chen Ning Yang and Robert L. Mills that was developed by, among others, Murray Gell-Mann, David Politzer, David Gross, Frank Wilczek, Yoichiro Nambu, and Kenneth Wilson from the United States; Gerard t'Hooft from Holland; and Kurt Symanzik from Germany.

The next revolutionary progress in our theoretical understanding of fundamental processes was the so-called electro-weak unification. There are four fundamental forces of nature known to us. First, gravity acts between all objects and governs the motion of things on earth and of other planets and stars. Second, electromagnetic forces between electrically charged bodies determine the structure of atoms and molecules, which are made up of electrically charged nuclei and electrons. These forces are responsible for the structure of matter that surrounds us here on earth. Third, the strong forces between the quarks, as described by quantum chromodynamics, determine the structure of atomic nuclei and their constituents. Finally, the weak forces are responsible for radioactive phenomena.

At first sight the weak interactions seem to be radically different from the electromagnetic ones, which bind electrons to atomic nuclei and atoms to molecules. Electricity is part of our daily life. The weak interactions are so weak that they do not bind particles to each other. They are observed *directly* by tiny scattering effects, when a neutrino passes through dense matter. The neutrino is a particle that is influenced only by the weak forces. For any other particle, however, the very small effects of the weak forces would be completely over-shadowed by the other forces.

The most important *indirect* manifestation of the weak forces is the radioactive decay of a neutron into a proton, accompanied by an emission of an electron and a neutrino. This is a very slow process. An isolated neutron takes an average of twelve minutes to decay, an interval that is immensely long compared to processes in the particle world. When the neutron is built into a nucleus, it usually does not decay at all. The change of structure and energy that the nucleus would suffer from a neutron changing into a proton does not allow the decay of the neutron in most cases.

The discovery that these two very different forces have a common origin came as something of a surprise to researchers. The reason for the different manifestations turned out to be what is called the "range of the forces." The electric force has an infinite range. It is true that the force between two charges falls off at larger distances between them, but the range is called infinite because the effect of the force is measurable even when they are far apart. The weak forces have a very short range; they are effective only at distances of less than $10^{-15}$ centimeters. This is less than a hundredth of the size of an atomic nucleus, so it is no wonder observing them is so hard.

The new electro-weak unification theory says that at those small distances, electric and weak phenomena are equally strong and rather similar. This is borne out by the observation that the weak inter-actions do indeed become stronger at smaller distances, although no one has yet been able to observe encounters of particles at the critical distance where they become equal to the electric forces.

If the force fields of the two interactions are equally strong at small distances, why do they differ so much at larger ones? The answer is connected to the "field quantum." In quantum theory every field is transmitted by means of field quanta. The field quantum of the electro-magnetic field is the light quantum, or photon, which is massless. It always moves with light velocity. When the field quantum of the weak interaction was discovered by Rubbia and collaborators in 1983, they found that those quanta have a very high mass, about a hundred times higher than that of the proton. Therefore, they cannot move far during the time of interaction, and this is the reason for the short range of the weak interaction. We cannot go into the exciting details of that theory here. Briefly, it showed that those two forces, which were regarded as different, are actually just components of one basic field. In this respect the theory can be compared to the great unification that occurred in the nineteenth century when James Maxwell showed that the electric and magnetic fields are only components of one basic field and that light is just a wave propagation of that field. Maxwell united electricity, magnetism, and optics. The authors of the present unification united electromagnetism and radioactivity. This theory was conceived in the late sixties by Steven Weinberg when he was at MIT and Sheldon Glashow at Harvard and independently by Abdus Salam and John C. Ward in England.

There were other interesting developments that aroused my interest. Astrophysics was rapidly growing in importance. A "cool" light radiation was found to permeate space. We call it "cool" because it corresponds to the heat radiation of matter at 2.7 degrees centigrade above absolute zero. At temperatures higher than the absolute zero, any material emits radiation. We feel it only if the temperature is high enough, as in the case of a glowing piece of iron. But sensitive instruments can detect it even at very low temperature. Most probably, cool radiation represents the remnants of the white-hot light that filled space shortly after the so-called big bang, which occurred about fifteen billion years ago. This radiation became cool because the expansion of the universe increases the wavelength of light in space. Thus, the hot light at the beginning was changed by the expansion into the currently observed cool radiation. It is something like an optical reverberation of the big bang still observed today.

Quasars and pulsars were also discovered during this unusually fertile decade. Much was learned about what may have happened during the first small fractions of a second after the big bang. Sometimes people are astonished that cosmologists speak of processes occurring in such infinitesimally short times after the beginning. Time intervals are relative. The time unit of human life on earth is a year because that is the amount of time it takes for the earth to return to its starting position in a voyage around the sun. The cosmic unit of time is a billion years. That is the timespan during which the cosmos today changes its character perceptibly—galaxies take billions of years to rotate. The atomic year is the time it takes for an electron to run around the nucleus: about $10^{-15}$ seconds (a millionth of a billionth of a second). The reasonable time unit of the subnuclear realm of elementary particles is again a billionth smaller. For those processes that had been going on shortly after the big bang, such short time intervals are what years are for us.

A great deal of activity occurred in most of the other fields of physics. New phenomena were found and new concepts developed about the behavior of matter under extreme conditions such as very low and high temperatures or very low and high pressures. New ideas had to be conceived about phase transitions (melting, evaporating) in order to explain the wealth of facts revealed by experiments made possible by the development of new technology. The invention of the laser permitted investigations of matter under irradiation by extreme light intensities. The interest of physicists turned to irregular phenomena such as turbulent flow and other manifestations of chaotic behavior that, up until then, had been largely neglected. Altogether, the great progress of physics in the sixties and seventies was linked to the development of better, larger, and more sensitive instruments and based on new theoretical ideas. To paraphrase Teilhard de Chardin, "The history of natural science can be summarized as the elaboration of ever more perfect eyes within a cosmos in which there is always something more to be seen."

All this and much more attracted my interest and curiosity. Although I often experienced the familiar joy of insight when I heard about or studied the new developments, I also found myself questioning my abilities. Was I capable of absorbing all of these new ideas? How much could I contribute myself? My doubts came from a perception I have that there are three periods in the life of a successful theoretical physicist. The first, when one is young, is the time of hard work on new ideas, either conceived by oneself, suggested by some master, or identified during collaboration with other young researchers. In the second period one has become reasonably well known and is invited to give talks at various universities and at conferences. One still contributes valuable original work, but more often with the help of younger colleagues or by suggesting a line of research to them. In the third period one receives invitations to give general talks surveying the whole field through the light of experience. A little later one is invited to deliver memorial speeches about deceased colleagues or talks on the philosophical and political aspects of science. Finally, all one is asked to do is give after-dinner speeches.

Thus, in the later periods the physicist becomes only a passive participant in research. One may still understand and enjoy some of the new ideas if they are not expressed in too-complicated formalisms. One keeps up with the developments by reading current publications, although that sometimes can be difficult because of the jargon, which changes today more quickly than ever. A much better method is to have it explained orally by the authors or by somebody who is well acquainted with the development. In that case one can ask for more detailed explanations in a more familiar language, an approach that obviously cannot be taken when reading.

I was lucky that Robert Jaffe, Kenneth Johnson, and Arthur Kerman, all professors at MIT, were always ready to explain recent developments in physics to me patiently whenever I asked them to do so. Many other colleagues in the various fields of physics were also ready to tell me about new ideas. Among these were Allan Guth in cosmology, Dan Kleppner in atomic physics, and Marc Kastner in condensed matter physics. They helped me remain more or less informed of what was going on.

At the time of my return from CERN, I was in the transition between the second and third period. I published a few research papers on problems related to the quark structure of elementary particles in collaboration with Roger Van Royen from Belgium, whom I had met at CERN and had invited to MIT. With the Israeli physicist Arnon Dar, I published several papers concerning the interaction of elementary particles with light and related problems. We used traditional quantum mechanics to understand the behavior of elementary particles as a combination of quarks. Julius Kuti, who was visiting MIT from Hungary, and I tried to explain some of the detailed features of the previously mentioned deep-inelastic scattering of electrons and neutrinos by protons.

There was one work of greater significance by a group of theoretical physicists at MIT to which I could contribute more than general encouragement. It involved one of the most striking consequences of the new quantum chromodynamics. The force between two quarks depends on their separation in a most unexpected way. Whereas the electromagnetic force is strongest when the electric charges are close together and it decreases with increasing separations, it is just the other way around in chromodynamics. The force is weakest at close distance and increases with growing distance! The first effect—the weakness of the force at close distances—carries the name "asymptotic freedom" because the particles feel no constraining forces when they are close together. The second effect, the increase of attraction at larger distance, is most unusual and surprising. This second effect prevents the quarks from getting out from the proton, because the farther they go, the more strongly they are pulled back. It was quite difficult to extricate an exact formulation of that increasing attraction from the theory.

We tried to simplify the problem in two ways. The strong increase of attraction at larger distances was schematically described by assuming that the quarks in the hadron are enclosed in a small spherical bag from which they were unable to escape. The weakened forces at small distances were replaced by assuming that the quarks move almost freely inside that bag. In other words, we described the forces between the quarks by a simple model, the so-called bag model. A model provides a simplified description of a complicated situation, retaining only its most important features. Its advantage was that one could calculate relatively simply the consequences of the quark structure of hadrons. It turned out, to our satisfaction, that this model exhibited properties close to the observed properties of hadrons. (In 1974, I collaborated with Alan Chodos, Robert L. Jaffe, Kenneth Johnson and Charles B. Thorn on a paper describing these results.)

We were not the only people who used this bag picture, but our paper and the subsequent papers by our group were the most extended treatments of that idea. Later our approach was referred to as the "MIT bag." It confirmed the quark structure of hadrons, in spite of its simplifying assumptions. Today, however, the bag model is slowly becoming obsolete because the computer allows researchers to carry out more accurate evaluations of quantum chromodynamics.

They largely confirmed the primitive conclusions from the bag model.

In 1967, I was asked to become the head of the physics department at MIT. Since there was no other candidate available at that time, I felt I could not refuse. The MIT physics department is the largest in the country. In addition to research in many fields such as astronomy, solid state physics, atomic physics, nuclear physics, and particle physics, it also is known for both general and specialized teaching. Every student at MIT has to take at least one general physics course.

Serving as the head of a large university department is quite different from being director of an institution like CERN. Compared to the researchers at CERN, the MIT professors are much more independent. They organize their own research, and the search for funding takes place in subgroups of the department. The head of a department has less direct influence on and responsibility for the research than does the director of CERN. However, the various teaching activities, except for very specialized courses, must be organized and assigned centrally by the department. When I took over, the machinery for teaching assignments in such a large department was already in the hands of an experienced staff. The duty of a newly assigned head was mostly to continue doing what had been successful before and to encourage changes and reforms only when necessary.

I knew from my CERN experience that I could not perform this job without the help of a person with whom I had excellent personal rapport and who would complement me by possessing qualities and experience I lacked. In other words, I was looking for another Mervyn Hine. In the first of my seven years as head of the department I was lucky to have persuaded Albert Hill, a physicist with experience in management, to work with me. Hill found an extremely efficient managing director in the person of Daniel Gould, whom we called "Piet." (Later, Anthony French took Hill's place. French had extensive experience in the teaching activities of the department.) With the effective help of these excellent men, who tried their best to protect me from too much administrative work, I was able to continue some of my research and could concentrate on establishing a healthy cooperative spirit in the department. The other important task was hiring the best people when positions opened up or when we were able to convince the university to create new positions.

Still, being the department head was not as rewarding an experience as my job at CERN. It was interesting to get in closer touch with fields of physics other than my specialties, but I found that I also had to attend countless meetings and conferences with university authorities. Many of these sessions proved the wisdom of the saying "A meeting is an event where minutes are kept and hours lost."

The late sixties and early seventies were good years for universities. Funding for science was still comparatively easy to get, and I was able to attract a number of excellent people to the department. When we needed an experimental physicist in particle physics, I thought of Sam Ting, a most promising young man. He had received offers from several universities, so I had to tempt him with a full professorship. Because he was only twenty-nine years old, I had a hard time convincing the president and other authorities that he deserved such a high position. I argued that he would probably win the Nobel Prize—a risky prediction that in time proved accurate.

Ting was famous for being totally committed to his work. He had planned and carried out important, successful experiments at the Brookhaven laboratory, at CERN, and at DESY (Deutsches Elektron Synchrotron), the German electron accelerator in Hamburg. It was astounding how many large projects he could carry out simultaneously. I believe he worked more intensely and with more care than many other experimental particle physicists I have met. When I told Ting that at MIT every professor was obliged to teach, he asked to be appointed as a research professor. I refused, but he still wiggled out of his teaching duties by using ingenious tricks such as letting some of his collaborators do the teaching. After his Nobel Prize the department let him do as he wished. He had a devoted group of collaborators, who worked extremely hard under his leadership, although it could not have been easy to work closely with him. He led his group with an iron hand, which bothered some people. One of his collaborators, Wit Busza, left the group and became a productive physicist on his own, making good use of the experience of having collaborated with Ting.

Ting has always been a very careful and prudent experimenter. He would not publish his results until he was quite sure that they were reliable. He kept his findings secret, forbidding his co-workers to talk about them until a series of control experiments had been performed. One of the most important results was the discovery of a new type of quark appearing in a particular meson, which Ting called the J-meson. (This was the $J/\Psi$-meson referred to in the previous chapter.) He was practically sure of the result just before the celebration of my retirement, which would have been a good occasion for announcing it. But he wanted to do some more checking. (Such reluctance is highly commendable. Too many apparently good results are published today that turn out later to have been wrong.) This time, however, another group at Stanford, under Burt Richter, discovered the same meson by another method before Ting decided he was ready to go public. As a consequence of his prudence, Ting had to share his Nobel Prize with Richter.

Our newly established Center for Theoretical Physics profited greatly from the appointment of Steven Weinberg. In 1966 he was a Loeb lecturer at Harvard, on leave from a full professorship at Berkeley. I had been impressed by his knowledge, originality, and inventiveness, and in 1969 I offered him a professorship at MIT, which he accepted because Harvard had not made him a similar offer. While at MIT, he worked out the unification of electromagnetic and weak interactions, work for which he later was awarded the Nobel Prize. He also wrote a book, *Gravitation and Cosmology,* which I consider the best introduction to Einstein's general theory of relativity for serious students. It was a pleasure to have him around, not only because of his deep understanding of theoretical physics but because of his wide cultural interests, particularly in the history of ideas.

After Weinberg became famous through his work at MIT, Harvard offered him a professorship. This upset me greatly because the two Cambridge institutions have an unwritten agreement not to raid one another's staffs. Unfortunately, Weinberg succumbed to the lure of a Harvard professorship and left us in 1973. However, divine justice intervened. After ten years he deserted Harvard for the University of Texas in Austin. It was rumored that he did so because of a lack of rapport with one of his colleagues.

With the addition of two theoretical particle physicists from Italy,

Gabriele Veneziano and Sergio Fubini, and a young American theorist, Roman Jackiw, we soon had one of the liveliest groups in the field. They joined Steven Weinberg, Herman Feshbach, Arthur Kerman, Francis Low, Kenneth Johnson, Robert Jaffe, and Kerson Huang to form a powerful group of active theoretical physicists. Ideas were exchanged in lively discussions. Research in theoretical physics requires a group of people working on similar problems. Only a very few theoretical physicists—Einstein is the most famous example—have developed their ideas alone. Most of the others need critical discussions with their peers to try out and develop new ideas. Our center vibrated with activity. Unfortunately, Veneziano and Fubini found it difficult to get accustomed to the American way of life, and after some fruitful years they left for CERN, where both of them now work.

In 1973, I resigned as head of the department. My good friend and collaborator Herman Feshbach took over. He had a harder time than I because the generous support granted to physical science from various sources dried up in the seventies, but he was able to maintain a high-quality department despite the difficulties. Herman had been a great asset to me all through my tenure at MIT. In the fifties we had worked together on some important questions related to the understanding of the structure of the atomic nucleus, as I mentioned earlier. When I came back from CERN, my interests were directed more toward particle physics, whereas Herman's remained with the physics of the nucleus. Nevertheless, we maintained an interest in each other's ideas, discussed our problems, and shared important suggestions, criticism, and advice. But more important, Herman and I developed a deep friendship and shared personal problems, departmental questions, and political ideas. Both of us were old-style liberals with similar worries about the developments in the United States and abroad. In difficult situations I sought Herman's advice first, and I always received support or, equally useful, sarcastic criticism. We both had an overdeveloped sense of humor, which made our conversations not only useful but amusing. It was quite natural that he succeeded me as head of the department. He is a more active and perhaps less contemplative person than I, but he has been a friend toward whom I could be open and genuine.

In addition, our diverse interests had a salutary effect on the center of theoretical physics at MIT. It became one of the very few institutions where nuclear and particle physics research went on in close contact, providing both endeavors with many advantages. I was always eager to bring these fields together. In the 1970s Amos de-Shalit and I initiated periodic conferences covering the two fields. These meetings were called PANIC, for Particle And Nuclear International Conferences.

In fall 1974, I reached the obligatory retirement age of sixty-five (recently raised to seventy, but regrettably too late for me), and I was transformed into a professor emeritus. In honor of my retirement a lavish two-day celebration was held, with colleagues attending from all around the world. (I find it strange that you "celebrate" the demotion to retirement.) There were many speeches about the present state of science and about political and philosophical questions. At the banquet I conducted a chamber orchestra and played the piano part of a Beethoven trio. The *New York Times* reported this illustrious gathering of international physics on the front page of the Sunday edition, and to my astonishment and great pleasure, the article prominently mentioned my musical contribution to the festivities.

After 1975, I no longer contributed to original research. I devoted my time to writing articles or giving talks in which I tried to explain in a simplified way some of the new ideas and concepts invented by others. I always considered such endeavors most rewarding for myself and, I hope, for my readers. It is regrettable that among scientists the presentation of ideas is not as highly valued an art as the creation of ideas. This is in stark contrast to music, where the performer is a partner equal to the composer. The outstanding artist of presentation in physics was the late Richard Feynman, one of the most impressive theoretical physicists of our time. He was great in every respect, both an innovative composer and imaginative performer, both brilliant in conceiving new ideas and inspiring as a teacher. He once said that if you cannot explain an idea in physics to a non-scientist, you haven't understood it yourself.

After my retirement, I definitely entered the third period in the life of a theoretical physicist. I was asked to give memorial talks for Cecil Powell and Wolfgang Gentner, both of whom, as I said earlier, had been active in the early days of CERN. I wrote articles about the frontiers and limits of science, and about art and science. I spoke at inaugurations of new accelerators—and at the closings of old accelerators—and I was invited to speak at several university commencement exercises. In 1985 the Public Broadcasting System's *Nova* series presented an hour-long program about my life and my convictions, entitled "The World According to Weisskopf," a title I fought bitterly but unsuccessfully. However, the program itself pleased me. I was particularly happy that it featured a segment in which my friend Dave Perlmann and I played Mozart and Beethoven sonatas.

When I realized that retirement was not going to mean an end to activity, I decided to undertake a project that kept me occupied for many years. I asked my old student Kurt Gottfried, now a professor at Cornell, to work with me on a book to be called *Concepts of Particle Physics*. After spending so many years studying, working, teaching, and organizing the physics of elementary particles, I wanted to write a book that emphasized an intuitive approach to our subject—not a textbook full of mathematical details. As we said in the preface,

> The goal of this book is to elucidate basic and well-established concepts of particle physics that students of this subject are expected to know. For that reason this book does not really belong to what we may call the written tradition of physics.
>
> Like every intellectual pursuit, physics has both a written and an oral tradition. Intuitive modes of thought, inference by analogy and other stratagems that are used in the effort to confront the unknown are transmitted from one generation of practitioners to the next by word of mouth. After the work of creation is over the results are recorded for posterity in a logically impeccable form, but in a language that is often opaque. The beginner is expected to absorb this written tradition and only the survivors of this trial by ordeal are admitted to circles where the oral tradition is current. We could only hope to strive toward our goal by leaning heavily on the oral tradition. This is not meant as an apology, for we believe that this tradition plays an essential role not

only in the creation of physics but also in the search for a deeper understanding. A wider dissemination of the oral tradition is therefore in order.

Some parts of the book were based on the lectures I gave at CERN during the summers after I left the directorship. However, I would not recommend it, in particular not the second volume, to readers not reasonably well acquainted with the basic concepts of quantum mechanics.

The collaboration with Gottfried was a pleasure, since his approach to understanding physics was similar to mine. Furthermore, it was instructive to me because he knew the mathematical formalism better than I. To our exasperation, however, neither of us could work exclusively on the book, given our other work and the inevitable interruptions. Intuitive reasonings are difficult to formulate, and we frequently had heated discussions about the best way to make a concept clear to the reader. After observing our method of working, Kurt's wife, Sorel, said, "You are not writing a treatise but a treaty."

My pedagogical interests were by no means confined to particle physics. Many other aspects of physics needed to be presented in a simpler and more transparent way than was usually done. My work in this direction was encouraged during a visit to our friend David Hawkins sometime in the late sixties. He was a professor of philosophy at the University of Colorado at Boulder. Our friendship began at Los Alamos, and we have seen each other frequently since then. I always found great pleasure in our discussions of the fundamental questions of science and philosophy.

A typical conversation took place during a hike in the Rocky Mountains around Boulder. I praised the power of quantum mechanics to explain the world around us. David responded by challenging me: "Can quantum mechanics explain why the mountains are usually several thousand meters higher than the valleys and not a hundred times more or a hundred times less?" I said, "Let's try to explain it together." We sat down on a fallen tree in a beautiful spot, found a pencil and some paper, and argued back and forth until we had some kind of answer based on a balance between the force of gravity and the strength of a material such as rock. The latter depends on the force between the atoms, which we estimated by using simple quantum mechanical arguments.

During another hike we sat at the shore of a small lake. The surface was calm, and we saw the mirror image of the surrounding mountain. Then a slight breeze started, and the surface darkened because small wavelets appeared. We asked ourselves, What is the wavelength of the waves that appear when the wind starts? Why are they of the order of an inch and not much larger or much smaller? (Of course, as the wind continues blowing, longer and also shorter waves occur.) We studied this phenomenon, and it turned out that it was also a balance between two forces. This time the balance is between gravity and the surface tension of water. The latter is connected with the interaction of water molecules. Dave quoted Tennyson describing, in *The Lady of Shalott,* the effect of a breeze on the surface of a lake:

> Willows whiten, aspens quiver,
> Little breezes dusk and shiver.

Another question was why raindrops are never larger in diameter than about a centimeter. Their size is similar to the wavelength of the first wavelets on a lake; it is also related to the size of droplets falling from a leaky ceiling. It finally turned out that the size of those droplets and wavelets is very close to the geometrical mean between the maximum height of mountains and the size of an atom. This is because the same kind of forces act in similar ways in all three examples.

Other questions had to do with the dimensions of atoms, with the nature of chemical forces that keep atoms bound in molecules, with the difference between liquids and solids, and with why the sun is as bright as it is. I always tried to be as simple as possible in the explanations. Sometimes I overstepped the limits of simplicity and should have remembered Pauli's famous remark, which I quoted earlier: "Simple it is, but it is also wrong." Certainly, I was not always able to follow mathematician and philosopher Alfred North Whitehead's sensible rule: "Seek simplicity, but distrust it." I did not work systematically on these problems. I prepared one or another subject for a talk or a series of lectures. Recently I began writing a book entitled *In Search of Simplicity* that will collect all these endeavors in a more systematic way. Herbert Bernstein of Hampshire College is my collaborator because he shares my penchant for simple explanations.

Explaining things as clearly and simply as possible is a task that has obsessed me all through my life as a scientist. I have tried to do this at all levels of sophistication, for learned specialists, students, and laypeople. My book *Knowledge and Wonder* and a number of articles and talks were directed at people with no scientific training. Too few scientists devote their efforts to such ends. That is why it was often hard for me to find people with whom I could talk about the problems and difficulties I encountered when trying to explain things in the best possible way. Richard Feynman would have been ready to discuss such problems any time I wanted to, but he lived far away in Pasadena. I was lucky, therefore, to have Edward Purcell nearby at Harvard. Ed was one of the few people who understood and valued such endeavors. He helped me enormously in clearing up vexing points and ambiguities. I always benefited from my talks with him about any kind of problem of that sort. Our approach to physics is very similar, although he is an experimental physicist and I am a theorist, so it was a pleasure to work with him.

Recently, browsing through my early writings, I found a paper entitled "Origin of the World According to Modern Research," written in 1922 when I was fourteen years old. Many of my facts were wrong, partly because I misunderstood some of the findings and also because cosmology was in its infancy at that time. But it was a clear, easily understandable presentation. Sixty-six years later I gave a popular talk on the same subject before the American Academy of Sciences, which was also published in the *New York Review of Books* in April 1989. I hope it was somewhat closer to the truth because of both the great progress in cosmology and my own improved understanding. But looking at my early talk and the more recent one, I was struck by the similarity in style. As early as 1922, probably under the influence of my tutor Hans Riehl, I was already emphasizing the emotional and inspirational aspects of scientific discoveries. As the ancient proverb says, "What is bred in the bone will not out of the flesh."

\*     \*     \*

I had been back in the United States for a time when I became aware of a serious lack of communication between the scientists involved in high-energy physics and the government agencies that provide the funds. At that time the Atomic Energy Commission (AEC) in Washington (and later the Department of Energy) was charged with regulating and funding these efforts. William Wallen-meyer was the able head of the division of the AEC in charge of high-energy physics. As noted earlier, while I was at CERN, we established the European Committee for Future Accelerators (ECFA), which contained representatives of the European high-energy physics community. Most of the managerial problems and the future plans of this field were discussed and presented to CERN and to the governments of the member states. After my experience with such issues at CERN, I was interested in seeing if such an organization would help solve similar problems in the United States.

In 1966, I proposed to Wallenmeyer that we organize an analogous American group. My proposal was accepted, and I was asked to be the chairman. (I was aware when I made the proposal of the built-in hazard that I might have to do much of the work.) We named the new organization the High Energy Physics Advisory Panel (HEPAP). Since its inception, HEPAP has been advising government agencies about the needs and future plans of the physics community and negotiating the amount of financial support that the government would be expected to supply. The idea was that such decisions should be thoroughly discussed between the government representatives and recipients before they were finalized. We hoped this would lead to a better understanding and greater personal contact between the phys-icists and the government agencies involved.

The first group of twelve to fifteen members Bill Wallenmeyer and I chose included the directors of the large government labora-tories, such as the Brookhaven National Laboratory and the Stanford Linear Accelerator Center. This turned out to be a mistake, because the directors were often more interested in the future of their own labs than in the progress of high-energy physics in the country in general. Later the directors were dropped. The problems we had to face were overwhelming and included such issues as the distribution of funds among the national labs, the amount of support to the universities where the users of the big labs prepared their experiments, the advisability of funding new expensive apparatus, and so on. All this depended, of course, on judgments about what line of research should be favored at the cost of others.

This brings up another problem that we had to face at CERN. Before the days of massive instrumentation, during the time of small science, individual senior scientists could choose their research program within the financial means available. In many other fields of science this is still true, but not in high-energy physics, where such decisions must now be made by committees. This is necessary because the same accelerators and many of the massive instruments are used by a large number of people. The choice and order of experiments must be carefully planned. Each laboratory has a committee charged with this task. Naturally, this setup leads to competition between different proponents and battles with the governmental agencies for more money. There were many such fights at the HEPAP meetings. The most difficult sessions were those at which the budget for the following year was discussed. Perhaps we could have avoided some of these arguments if we had had something akin to the Bannier plan of CERN where the budget was planned ahead for several years.

In spite of the difficulties I enjoyed these meetings. They were also interesting from a scientific point of view because some of them took place at the laboratories in question, where we were also able to learn about the ongoing experiments. As chairman, I could use my CERN experience to settle disputes between competing groups, and Bill Wallenmeyer and I had an excellent rapport, which helped when we were negotiating for financial support from the government. I once told him that I found it easier to deal with twelve governments than with one. At CERN we could play one government against the other. Often it was helpful in an argument with country X to point out that country Y had no objections to this or that. But at HEPAP we had only a single source of support.

In 1976, I was asked to become president of the American Academy of Arts and Sciences, a national institution with headquarters in Boston. I accepted with pleasure. This was just the sort of thing I wanted to do now that my retirement had freed me for such activities. The academy was founded in 1780 by John Adams as a "learned society to cultivate every art and science which may tend to advance the interest, honor, dignity and happiness of a free, independent and virtuous people." At present it has about 3,000 members, all distinguished scientists, engineers, humanists, artists, and public servants, as well as some businesspeople and industrialists. In contrast to the National Academy of Sciences in Washington, it is completely independent and gets no financial support from the government except for an occasional special project.

The academy holds monthly meetings in its Cambridge house and less frequent meetings at its other centers in the Midwest and on the West Coast. Usually the meetings feature an address on a subject of topical interest by some outstanding scholar. In addition, there are periodic study groups on issues in a wide variety of intellectual activities, with special emphasis on critical issues facing society. The findings of the study groups are published in the academy's journal, *Daedalus,* which has a national circulation, or in books published separately. The problems of arms control and international relations receive special attention, and the academy's independence from the government makes it possible for the study groups to evaluate even those ideas and proposals that are opposed to the current national policies. Addressing these issues within the academy was one of the reasons I was especially interested in getting involved in its activities.

I was also attracted by the opportunities for getting in touch with people not involved in natural science, especially those in the humanities. In my farewell talk at the end of my three-year term as president, I gave credit to my contacts with historians, sociologists, and artists for broadening my views. I also remarked that one of the things I had learned was that whereas scientists use only one-third of the words necessary to make themselves understood, humanists use three times as many.

During my term as president we planned a new building for the academy, an activity I thoroughly enjoyed. Edwin Land, my old friend from the Friday Night Supper Club in Cambridge, had given us a large sum for the building, and Harvard had donated a lot in a pleasant wooded area of Cambridge adjacent to the campus. Our first job was to interview prospective architects. Most of them showed us photographs or drawings of buildings they had designed and told us about how they worked. Few of them wanted to hear our ideas about what we considered an appropriate house for the academy. After a time we found the firm of Kallman, McKinnel and Wood, which seemed to be willing to hear what we had to say. It was a special pleasure to discuss the details with this firm's staff and to make suggestions, most of which they readily accepted. We wanted a building that was at once dignified and intimate, in harmony with the character of the academy. Their design fit well into the surroundings and provided us with a house that included a modern auditorium for lectures, as well as comfortable and elegant areas for our social activities. It became an extraordinarily pleasant place. The only drawback was that, as usual, the costs exceeded the original estimates.

My immediate successor as president was legal scholar Milton Katz, but after his term was over, Herman Feshbach assumed the post. Herman applied his usual creative drive to the job and introduced a number of innovations such as regular weekly lunches for local members and a regular Saturday afternoon series of concerts or talks of general interest.

During this period I was also dreaming of an international laboratory. My experience at CERN had led to this idea. I thought it made sense that the next very big accelerator (VBA) be built cooperatively by all nations interested in such basic questions as the fundamental structure of matter and the behavior of matter at very high energies. Such a world laboratory would be a symbol for science as a bond between all humanity, just as CERN was a symbol for the united nations of Europe. I was not alone in having this vision. The other active dreamers were two Americans, Leon Lederman and Robert R. Wilson. Both are very much involved in particle physics, mostly at Fermilab (a large laboratory in the neighborhood of Chicago), where Lederman had succeeded Wilson as director, and Wilson had originally directed the construction of the laboratory. We tried to convince the leading particle physicists in the United States, Western Europe, the Soviet Union, and Japan to think seriously about a VBA to be constructed through the collaboration of many countries and eventually utilized by them cooperatively—in other words, a CERN on a world scale.

We proposed the idea, and it was discussed at meetings of ICFA, the International Committee for Future Accelerators. ICFA's members, the world's leading high-energy physicists, made suggestions about what kind of new accelerators should be constructed, as well as where and when. The committee also tried to regulate the access of researchers of the various nations to existing or future national or regional accelerators. During the next nine years I attended four meetings at which the VBA was discussed. They were held in Vienna; Morges, Switzerland; Serpukhov, USSR; and New Orleans. We always had a good number of Soviets participating. I addressed several meetings of ICFA, urging a world approach to the planning of new machines without recourse to narrow national and regional interests. When my health prevented my traveling to Tokyo for one of these meetings, I made a video cassette of my talk and introduced it by saying that because I was unable to bring the atoms of my body to Tokyo, they would have to be content with the light and sound emitted by those atoms.

The world machine, however, has as yet remained just a dream, for several reasons. First was abiding self-interest. The four main instigators—the United States, Western Europe, the Soviet Union, and Japan—always had one new accelerator planned or under construction. They did not want to endanger the financial support of their governments for their pet national projects by proposing a world machine. Furthermore, people were afraid of the heavy bureaucracy that an international laboratory would involve, particularly with Soviet participation. There was a general feeling that an accelerator is best constructed nationally or regionally, as in the case of Western Europe. The international character of the science could be better served, they argued, by an open-door policy allowing teams from other nations or regions to use those machines by, for example, international participation in the committees that decide their use.

275

Such arrangements are, in fact, common, but the guest researchers from outside still do not have quite the same rights as the national or regional users.

In recent years the idea of an international accelerator lab has vanished from the discussion. All four protagonists are planning to build new and bigger machines on their own. Unfortunately, this planning is not well coordinated. Every region wants to be first with some higher energy. Nationalistic pride begins to play a role and displaces rational discussions on how best to use the different possibilities around the world without duplicating efforts. My hope is that the high cost of such new installations will force the communities of the different regions to adjust their future plans to the needs of a successful search by the world community of physicists into the innermost structure of matter.

In 1983, I received an invitation to the White House for what was announced as an important talk by President Reagan of concern to the scientific community. I accepted, of course, and found a number of colleagues and friends assembled in the afternoon, together with a lot of high-ranking military brass. The president welcomed us and said that some of his aides would explain the content of his speech, which was to be delivered in the evening. Then he left. The aides included Robert McFarlane, head of the National Security Advisory Board; Fred Ikle, then undersecretary of defense; and Jay Keyworth, the Presidential Science Advisor. They told us that the president was asking the scientific community once more to save the country, as it had when it constructed the nuclear bomb. The new project was a shield that would prevent enemy missiles from reaching the territory of the United States and its allies by destroying such missiles on their way through space with powerful lasers or other penetrating means. It would make nuclear weapons impotent and obsolete. They were telling us for the first time about the strategic defense initiative (SDI, or more popularly, Star Wars).

This announcement astonished us. There was a short discussion in which Edward Teller strongly supported the new weapons defense system. (He may have talked Reagan into this romantic proposal in the first place.) But at least four people spoke sharply against it:

McGeorge Bundy, Simon Ramo, Hans Bethe, and I. We used two arguments. One described the technical difficulties, which seemed to most of us practically insurmountable. The second was that SDI would force the other side to increase its missile stockpile to overcome the losses. Even if the project were technically feasible, it would be a serious obstacle to arms control. At the cocktail hour Teller reproached Bethe and me for being too pessimistic about the possibilities of this new project.

The president finally gave his speech, not in person, but on a television monitor after we had eaten dinner. Since that evening the government has spent an inordinate amount of money on the SDI project, which turned out to be beset with many presumably unsolvable difficulties. Furthermore, the insistence about proceeding with SDI indeed made it much more difficult to get an agreement with the Soviet Union on reducing the number of strategic weapons. After that meeting at the White House, the major focus of the peace efforts of the organizations to which I belonged were devoted to convincing the public of the negative effects of pursuing SDI.

During my tenure at CERN, I had stopped speaking or writing about the problems and danger of the nuclear arms race, because I considered it inappropriate for an official of an international organization to do so. But the situation had become worse every year. When I left CERN and returned to Cambridge, I wanted again to join efforts to establish a better world. During my frequent visits to Europe I tried to get in closer touch with people who were also interested in an understanding between East and West. I frequently visited Bruno Kreisky, who was chancellor of Austria from 1970 to 1983.

I admired Kreisky's thorough understanding of foreign policy in the East and West and his role in the non-Communist Socialist movement. Under his leadership, Austria had recuperated economically from the bad years of the Nazi occupation and had become a prosperous country. Although he was of Jewish descent, Kreisky was reelected several times, each time with a larger majority, an amazing fact in a country by no means free of anti-Semitism.

In the 1970s I was with Kreisky in Klagenfurt, a small city in southern Austria, near the Yugoslav frontier. There, I witnessed an

277

event that illustrates his personal courage and integrity. At that time there was trouble with neo-Nazis who were opposed to the Croats living near the frontier. The Nazis particularly took offense at the street signs and other announcements written in both German and Croat. A group of these hoodlums had torn down many of the Croat signs. When Kreisky arrived at Klagenfurt, the Nazis demonstrated on the square in front of the town hall, where he gave a rousing speech attacking them. After his talk the police suggested that, for his safety, Kreisky should leave the hall by the back door. Kreisky refused. "An Austrian chancellor does not leave by the back door, (*beim Hintertuerl*)," he said, loudly enough for the journalists to hear.

As the head of the Austrian government Kreisky was often accused of being too friendly to the Arabs. I think he knew better than many others how to deal with them. After Arabs attacked a building where Jews emigrating from Russia were housed, he closed the building in an action that was strongly criticized. But he was able to care for twice as many immigrants in safety, spreading them among several different places. He negotiated with Arafat long before other politicians recognized that peace in Israel could only be established with Arafat's cooperation, and for that he was also criticized.

I felt it was my duty to join the scientists in the United States who continued to warn the public that the two superpowers were on a collision course. I gave talks and wrote articles on that subject and supported groups of citizens devoted to the same aims, such as the movement to freeze nuclear armaments. These groups represented only a minority, although the freeze movement had large public support for a short time. The general public and their representatives did not worry much about the arms race. On the contrary, they were impressed by the deterrent effect of nuclear weapons, which had prevented a war between the superpowers. The great powers found no other way to maintain this state of affairs but to increase and improve their nuclear arsenals. They had reached an insane level of nuclear explosives, equivalent to six thousand times the amount used in the Second World War.

Fear and distrust had driven the superpowers into this situation, giving both sides about fifty times the capacity to destroy the other completely. There was fear that the other side might risk a first strike, fear of being behind in certain weapon types, and distrust manifested by a perception that the other side was bent on world domination. Each side accused the other of acts of aggression. But in many instances, acts perceived to be aggressive by one side were considered defensive by the other. For example, we pointed out, what the United States sees as aggression by the Soviets in Afghanistan and Southeast Asia may be considered by the Soviets to be defensive acts meant to insure the security of their borders or to counterbalance Western or non-Communist power. Conversely, U.S. actions to prevent the spread of Communist influence in Central America, the Middle East, and Vietnam may be perceived by the Soviet Union as attempts to increase Western influence wherever possible. We argued that a better understanding between East and West was necessary to achieve effective arms control and a reduction of nuclear weapons. In spite of our efforts, the scientific community was, on the whole, unable to convince the public of the necessity of an understanding with the Soviets in order to reduce the insane proliferation of nuclear weapons.

My friends and I persisted in our efforts to maintain collaboration between the United States and the Soviet Union even after the Soviets invaded Czechoslovakia and Afghanistan. It seemed to us that in times of stress it was even more important to maintain relations with scientific circles, most of whose members, after all, disapproved of their government's oppressive policies, even if they were not free to express these opinions. I am glad to report that the leadership of the large U.S. laboratories continued to admit Soviet scientists.

During a visit to Moscow in 1976, I was asked to see the vice-president and the secretary in charge of physics of the Soviet Academy of Science. They told me of their intention to offer me a foreign membership in the academy—probably because of my efforts to bring about a fruitful collaboration with Soviet scientists—and asked whether I would accept. I answered that I would be delighted to join the academy, since I am always in favor of international scientific relations. But I expressed one reservation: I would immediately resign with a public statement if and when Andrei Sakharov was expelled from the academy. I asked them to retract the offer if they could not agree to this condition. This was during the time when

Sakharov was being severely attacked by the government, although he had not yet been exiled in Gorky. I knew that these gentlemen had signed a statement condemning Sakharov's political activities. They assured me that I was completely free to resign whenever I wished. My reservation was no reason whatsoever to retract their offer. So I became a foreign member, in some ways perhaps balancing my membership in the Pontifical Academy, which started in the same year.

Shortly after that conversation I went to visit Peter Kapitza, with whom I could talk quite openly and frankly. I told him what I had said, and he approved. He remarked that most academy members sympathized with Sakharov, and many of those who had signed the statement against him had done so out of political expediency. Of course, Kapitza did not sign the statement. He implied that my stand might even help the academy resist government pressure to expel Sakharov. In fact, Sakharov remained an academy member until his death.

I enjoyed the privileges of academy membership in my subsequent visits to the Soviet Union. Visiting members stay in the academy's comfortable guest house, and a car and driver are put at one's disposal. The first time I used this service, I asked the driver to take me to 48 Ulitza Chkalova, Andrei Sakharov's house. I knew Sakharov from several scientific conferences and admired him both for his role as a pioneer in modern cosmology and as a courageous fighter for a better world. I told the driver to wait for me.

After I had visited with Sakharov and his wife, Elena Bonner, for two hours, Andrei brought me down to the street. He embraced me before I entered the car. As we started off, the driver, obviously a KGB man, said, "You should know that Sakharov is a very bad man." I replied, "How can you say that? He is a dear friend of mine and a very great scientist." He admonished me, "You should use your influence to make him stay with his science and not dabble into politics where he does immense harm." I expressed my complete disagreement and my conviction that his political ideas were of great significance and should be followed. I don't remember everything I said, but I am sure it was duly reported word for word and can be found in the KGB's files on the academy.

A change in U.S. public opinion about nuclear war began slowly to take shape only in the eighties. It was caused not so much by the efforts of scientists but by other groups, such as International Physicians for the Prevention of Nuclear War and the pastoral letter of the American Catholic bishops. It seemed the public believed physicians and religious leaders more than scientists. However, the decisive event that caused a turn of public opinion in the Western world was the fundamental change in the attitude of the Soviet Union under the leadership of Mikhail Gorbachev. Before his rise to power the Soviets took a reluctant and distrustful attitude toward any arms control measures. Suddenly Gorbachev initiated new proposals and unilateral measures to reduce weapons and improve East–West relations. Quickly the roles of East and West were reversed. Now the West was reluctant to go as far toward easing tensions and reducing weapons as the Soviets proposed. However, some positive results have occurred. The intermediate-range nuclear missiles were eliminated from Europe, and as I write, much is changing for the better in Eastern Europe.

Sakharov was vindicated at the end of his life. The changes in the Soviet Union under Gorbachev go back to his ideas. I doubt they would have occurred without Sakharov's persistence and courage. Like Moses, he led his people to a better world and died before they arrived. Like Moses, he had to fight golden calves that threatened the fulfillment of the promise. Will the Russian people arrive at the land of milk and honey? We hope so but cannot be sure. His spirit is no longer present to advise and guide them.

Perhaps after forty-five years of cold war there is finally reason to be more optimistic. The wind has turned in world affairs. It blows from another direction, albeit still only weakly; there are threatening clouds on the horizon that may change the direction again. But I am grateful that I have lived to see our efforts to make this a more peaceful world seem to bear fruit.

# THIRTEEN

⊠

# Working with the Pope for Peace

IN 1976, I was given an opportunity to address matters of disarmament in a new and important arena when I was elected to the Pontifical Academy of Science, an advisory body to the pope. The letter from the Vatican announcing my election was something of a surprise, although I had some inkling that it was in the making. I believed I had been asked to join this illustrious body through a recommendation from Louis Leprince-Ringuet, with whom I had developed a close friendship during my stay in Paris and his subsequent involvement in CERN. He had been influential in getting the French government and the French physics community to support CERN, and we had visited him often in France, particularly in his summer home, an old chateau in Courcelles in Burgundy. A devout Catholic, Leprince-Ringuet had been a fellow of the Pontifical Academy for some time and a member of its governing council. The president of the academy, Carlos Chagas, also knew me and had heard about my interest in science and social affairs. But election to the academy is by secret ballot, so I will never know the full story.

The Pontifical Academy was established to advise the Holy See in matters of science. Membership has never been restricted to Catholics. Among the seventy members there when I began my service were about fourteen or fifteen Jews, numerous Protestants, and one or two Muslims. Indeed, Catholics were in the minority, making up only about 30 percent of the membership. However, the president of the academy is always a Catholic, as are the four members of the governing council, who serve with the president. Today there are four female members of the academy: Rita Levi-Montalcini, a Nobel Prize–winning biologist from Italy; Maxine Singer, an American molecular biologist; Johanna Dubereiner, an agronomist from Brazil; and the American biologist Beatrice Mintz. In particular, I greatly valued the chance to be together with Rita. We shared many interests in human questions, particularly the problems facing Israel. Our attitudes toward our Jewishness was also rather similar, as she came from an Italian Jewish family that was assimilated and resembled mine.

The academy was founded in 1603 and given the name "Academia dei Lincei"—the Academy of Lynxes, after the animals known for their keen eyesight and intelligence. Galileo was one of its founding members, but when he was excommunicated, he got no help from the academy members, who caved in to church pressure and did not support him. Until 1847, when the academy was brought under the supervision of the Vatican by Pope Pius IX, it barely survived and functioned hardly at all. It reemerged in its present form in 1936, when Pope Pius XI revived it and began to admit members of all religious beliefs. The academy was then no longer known by its zoological name, which had been taken over by the Italian National Academy of Sciences. However, membership was actually only honorific, and the academy existed as an ineffectual body until the early seventies, when it received its current mandate to help the pope formulate statements on scientific matters.

Some have attributed this change from slumber to action to the dynamic leadership of Carlos Chagas, a highly respected and personable Brazilian biomedical scholar who was named president of the academy in 1972 by Pope Paul VI. When he was chosen, he told the pope, "What I want is quite clear. I would like to see the academy change from a body of prestige to one of action." He immediately recruited new members, including several from Communist countries. Chagas established an equally good personal rapport with Pope John Paul II, who shortly after his elevation to the papacy expressed an interest in science and its relation to social problems. We were very impressed when, early in his reign, John Paul came to one of our meetings—he came to us, not we to him—to tell us how much he valued our advice.

The work of the academy is carried out in plenary sessions held every two years and through conferences on special topics. A general theme is chosen for the plenary sessions, and a few people are asked to give talks, which lead to much discussion among the members. Examples of such general themes are "The Responsibilities of Scientists," "Basic and Applied Science," or "Science and Society." Direct discussions of the relations between religion and science have always been avoided.

The academy also organizes study weeks on particular scientific subjects chosen by the council. Most of these proposals are made by the president, who must submit them to the pope for approval. The study weeks are organized and presided over by one or several members of the academy. The best scientific experts in the field are invited from all over the world to participate, whether they are members or not. These weeks are interesting events from a scientific point of view as well as from a social one. Usually the pope addresses the participants at a reception that he hosts.

The academy's main interest is in fundamental scientific questions and problems of the relation of science to society. The topics of the study weeks reflect that focus. Some deal with purely scientific subjects, such as new ideas about the origin of the universe, large-scale motion in the universe, and theories of evolution. Others, however, are directed at practical subjects of applied science, such as problems of energy, molecular mechanisms of carcinogenic activities, modern methods of agriculture, genetic engineering, or medical problems such as the treatment of retarded children and other related subjects. Final reports are published containing the proceedings of the study weeks. These important documents command respect in the scientific world. As an example, the report on evolution contained the following statement: "We are convinced that masses of evidence render the application of the concept of evolution to man and the other primates beyond serious dispute." The pope's approval of this document shows how much the Catholic church distances itself from the so-called creationists.

Unfortunately, the important topics of birth control and abortion have been strictly out of bounds. Obviously, a considerable number of academy members, including myself, disagreed totally with the

Catholic church's views on these issues. We had to shun these topics and concentrate on issues where we could expect support, such as the fight against nuclear war and, recently, the protection of nature against the effects of harmful human actions.

The academy is housed in a wonderful little Renaissance pavilion in the private Vatican Gardens. The building was erected in 1562 by the Medici pope, Pius IV, and is an exquisite architectural gem, its pleasant halls decorated with opulent ceiling frescoes in the style of the High Renaissance that sometimes seem rather audacious for a Vatican building. During a working session the American astronomer Martin Davis once said, "If you get tired of the talks, you can always look up at the dancing cupids."

A terrace is in front of the building, and at the opposite end of the terrace is a portico from which one is afforded a breathtaking view toward St. Peter's Cathedral and over the garden. It is easy to picture Pope Pius IV and his colorful company gathered around that table admiring the view while partaking of lavish meals and the wonderful local wine. The pavilion was enlarged for the academy. An assembly hall in a more modest style was added without destroying the charm of the original building.

We were always invited to bring our spouses to the meetings, and Ellen and I soon discovered that the person in charge of the Vatican library, Monseigneur Alois Stickler, was an Austrian who was happy to show us the illuminated manuscripts from the Middle Ages and the many other extraordinary objects these collections contain. I think he also enjoyed the opportunity of speaking German with someone who shared his Austrian accent. Thanks to him we had access to art treasures that ordinary visitors to the Vatican do not see.

It happened that one of the Academy's plenary sessions coincided with the election of Pope John Paul II. We were all invited to the great mass in St. Peter's Cathedral that was celebrated by the Congress of Cardinals.

The evening before, we were sitting in the hotel with Alexander Rich and his wife, Jane, who are close friends from Cambridge. Like me, Alexander is of Jewish origin. Before we went to our rooms, I said, "See you tomorrow at mass." He replied jokingly, "Would you have ever dreamed that we'd say this to one another?" From the middle of the church, where we were seated for the mass, we had a wonderful view of the cardinals from all over the world as they passed by in their splendid robes. I pointed out one who looked more impressive than the others and had a certain resemblance to my brother. This turned out to be Cardinal Karol Wojtyla, from Cracow, who was elected as the new pope a few hours later. It was an unforgettable event.

My first activity at the academy was in connection with the celebration of Einstein's hundredth birthday in 1979. It was the first time a pope had presided over festivities honoring a Jew. The scheduled speakers included President Chagas, the famous British physicist P. A. M. Dirac, and myself. Chagas presented the life of Einstein, and Dirac spoke in complicated terms about Einstein's influence on contemporary science. I took it upon myself to explain what Dirac had meant to say. But to everyone's surprise, the pope used the Einstein celebration to express his views on the relation between religion and science. In a remarkable change of church policy he stated that the church's judgment against Galileo was a mistake. The true scientist, he said, searches for a truth that the church cannot deny. Since the universe was created by God, scientific truth can never be opposed to religion. In interpreting the Bible one must take into account the cultural background of the times during which it was written. For this reason, the contents of the Bible should be interpreted symbolically rather than literally.

He went on to say,

Today man always seems threatened by what he creates. . . . [This is] the central chapter in the drama of human existence. . . . The domination of man over the visible world which is given him as a task by the Creator himself, should be based upon the priority of ethics over technology, of the preeminence of people over things and the superiority of the mind over matter.*

Quoting a statement of the Vatican Council, he said, "Methodical research, in all areas of knowledge, if it is carried out in a truly

*Commemoration of Albert Einstein (Citta del Vaticano, 1979).

scientific manner and if it follows accepted moral standards will never really be in opposition to the faith." Finally, he noted that in the academy believers and nonbelievers work together, uniting their search for scientific truth. He then quoted the physicist Abbé Georges-Henri Lemaître, famous for his research on cosmology and the determination of the age of the earth:

> Both [the scientist who is a believer and the scientist who is a nonbeliever] strive to decipher the parchments on which nature has written its secrets and where various stages of the long evolution of the world have left confusing and intermingling traces. Perhaps the believer has the advantage of knowing that the enigma has a solution, it is in the final analysis the work of an intelligent being; so the problems posed by nature are here to be solved, and the degree of difficulty is without doubt appropriate to the present or future intellectual capacity of humanity. This will perhaps not give him greater resources for his investigations, but will help to support his feeling of healthy optimism without which a sustained effort cannot be maintained.

The pope then added, "I wish all of you this healthy optimism of which Abbé Lemaître speaks; it is an optimism which takes its mysterious but very real origin from God in which you have placed your faith or from the unknown God to which the truth, the goal of your enlightened research, reached out."

At the time of preparation for the Einstein celebration I was in our summer house near Geneva. My friend and former student Herbert York was in Geneva as U.S. ambassador to negotiate a comprehensive test-ban treaty to halt all bomb testing, including underground tests, which were not banned by the partial treaty between the United States and the Soviet Union. Unfortunately, these negotiations did not succeed. When I told York what was going on at the Pontifical Academy, the suggestion was made by one of us— I no longer know which one—to use the academy as a platform for the pope to speak out against nuclear war and nuclear weapons.

President Chagas, to whom I transmitted this idea, was very enthusiastic. On my recommendation he called in a few people, including York, to discuss this subject and to see what we might propose for a declaration by the pope. We composed a memorandum emphasizing the dangers of a nuclear war, its consequences for the population and the environment, and how terribly destructive a nuclear war would be even if only a few nuclear weapons were used. Chagas submitted this memorandum to the pope. His Holiness must have been impressed by it. He used our phraseology in several talks he subsequently gave. In his homily delivered at the beginning of the new year in 1980, he actually used full phrases from our memorandum and said that this was what *his scientists* had told him. He spoke out strongly against the nuclear armaments race and against the increasing number of nuclear bombs.

Our working groups on the danger of nuclear war met at irregular intervals during the subsequent years, with different participants. We invited medical experts, who emphasized the difficulties of caring for the large number of victims when most hospitals would be destroyed and medical help practically unavailable. Some of the papers we published were used by the United Nations as working documents. Our memoranda may also have had an influence—of course I cannot be sure about it—on the pope's decision to visit Hiroshima, where he gave a talk along the same lines in 1981. I was interviewed by a *New York Times* reporter who said he was told that I had inspired the pope to speak out against the nuclear arms race. I answered jokingly that the pope is inspired by God and not by a Viennese Jew. I found this statement published in the *Times* the next day.

In fall 1981, just before a meeting of U.S. and Soviet delegations in Geneva on the reduction of nuclear weapons, the pope wrote a letter to the heads of state of each of the countries possessing nuclear weapons, emphasizing the dangers of nuclear war and asking them to receive a delegation of the academy to discuss these issues. Delegations were dispatched to Moscow, Paris, London, and Washington, and to the United Nations. I was a member of the delegations sent to Washington and to the United Nations. The group sent to the White House also included David Baltimore, the MIT Nobel laureate in biology, Howard Hiatt of the Harvard Medical School, and Marshall Nirenberg of the National Institutes of Health. We were accompanied by the apostolic delegate to the United States, Archbishop Pio Laghi.

Members of the press were informed about our appointment, and we asked them to be in the White House press room afterward so we could report on our conversation with the president. We were not surprised that we were kept waiting for some time in an anteroom after our arrival, but when a navy captain led us to the Oval Office, where the president greeted us, we thought it very strange that he didn't even ask us to sit down, nor did he sit. As the leader of the delegation, I asked him whether he had seen the letter from the pope. He looked quite taken aback and said, "What letter?" Obviously, he had not been prepared for the visit. I tried to sum up the contents of the letter, and then each of us spoke briefly about the dangers of nuclear war.

This occurred just a few months after the assassination attempt on President Reagan. Dr. Hiatt pointed out that the president's life had been saved by good medical care, but that such medical care would not be readily available during a nuclear war. Perhaps these remarks finally reached Mr. Reagan, because for the first time he seemed to show some interest. He agreed that prevention of nuclear war was extremely important, but he contended that negotiations must be carried out from a position of strength because the Soviets were stronger than the United States. We had often heard this kind of statement from him.

After about ten minutes he showed us out again. We asked the navy officer where the press room was, but we were told that nobody was there. He said the journalists were probably waiting outside the entrance. As it was a very rainy day, this was improbable. Indeed, nobody was there, so we didn't know what to do. We had left the White House, and it was difficult to go back in again. None of us knew how to find the press room. The assistant clergyman who accompanied the archbishop then spoke with one of the guards and was able to convince him that we had to get back in. I have always thought that this guard was Catholic, because he not only let us in but also directed us to the press room. Several dozen journalists were waiting impatiently for us. The navy captain was also there. When I tried to reproach him, he denied having told us that there was no one in the press room. When we spoke to the press about our talk with the president, we mentioned our cavalier treatment, and all of this was carried in many newspapers the next day. However, we were crowded off the front pages by the news from Poland about the crackdown on Solidarity.

A year later President Chagas, with the pope's approval, invited the presidents or representatives from all the academies of science of the world to come to Rome to issue a statement about the dangers of nuclear war. Of course, I was delighted about this initiative and offered my help in preparing the statement. A small group of us met several times in different European capitals to formulate a preliminary draft.

The representatives of the academies assembled in Rome in September 1982. We were pleasantly surprised by how many academies' representatives accepted the invitation and pleased that even those from the Soviet Union and other Communist academies came. The discussions were sometimes difficult because we had to overcome the natural reluctance of academies of science to go too far in any political action and the opposition of the Communist academies to religious terminologies.

The statement contained an appeal to religious leaders that provoked some protest among the Eastern participants until I suggested adding the phrase "and other custodians of moral principles" in place of more formal religious nomenclature. There were a few other changes in the original draft, but the final version was approved unanimously, and the pope endorsed it wholeheartedly. The declaration points out the irreparable effects of a major nuclear conflict; warns of the high probability that a nonnuclear war between powers possessing nuclear weapons may escalate into an increasingly devastating nuclear war, and refers to the deterioration of the world situation due to the growth of mistrust and suspicion and to the ever-increasing arms race. It stresses the absolute necessity of steps toward changing those trends and recommends preventing nuclear catastrophe by stopping the arms race, reducing the number of nuclear missiles, and negotiating arrangements to stabilize the military and political situation. It contains an appeal to all nations to refrain from the first use of nuclear weapons, an appeal that is most appropriate for a document sponsored by a religious organization. In the effort to avoid war and achieve a meaningful peace, it states, the force of ethics and moral conviction are as vital as the powers of intelligence.

This important warning coming from world science under the auspices of the Vatican was widely reported in the international press.

I believe that the explicit condemnation of the arms race by the Pontifical Academy influenced the American bishops, whose pastoral letter in 1983 was critical of the arms race and disapproved of the policy of nuclear deterrence as anything more than a temporary measure, on the grounds that it threatens civilian populations. The bishops urged the intensification of negotiations for arms control. This pastoral statement had an important influence on public opinion toward the recognition that nuclear war was not a viable alternative.

In the 1980s scientists began to be concerned about the possible reduction of the earth's solar heat from the dust and smoke of a nuclear conflagration. This so-called nuclear winter was on the pope's mind, too, and in 1984 he asked the academy to study the dangers of this phenomenon. The following year we were asked to address our attention to the proposed strategic defense initiative. We issued a document emphasizing its enormous technical difficulties and the negative effects the proposed defensive measures would have on efforts to reduce the number of nuclear missiles possessed by both sides. The Vatican never released our document, perhaps because it was more political than our other statements. There was a certain resistance in the Curia, the governing body of the Vatican, to academy statements of direct political consequence.

The members of the academy were and are people with a variety of ideas on many issues. At plenary meetings we often discussed how far the academy should go in dealing with political issues such as the arms race, SDI, or nuclear winter. Not all the members agreed with the activities of our subgroups on these issues. Furthermore, we had heard rumors that the U.S. government had urged the Vatican not to go too far in the area of politically sensitive subjects. Supposedly, the American delegate to the Vatican, a strong supporter of President Reagan, had warned the Vatican that President Chagas and I were too radical and one-sided on these issues. I am not sure the rumors were correct, but I would have been glad to be attacked from that direction.

In the course of my work in the academy I had many occasions to meet the pope personally. I was always impressed by his thorough grasp of the subjects under discussion and by his sympathetic attention to the opinions of others. Altogether, he is a friendly, outgoing person who does not use his authoritative position to intimidate his partners in a discussion. Of course, we thoroughly disagreed on many subjects, such as feminism, abortion, and birth control. I found it hopeless to discuss these issues with the pope, so we carefully avoided them in our personal encounters.

One of the most positive aspects of my connection with the Pontifical Academy was a growing friendship with President Chagas. I became increasingly impressed by his wisdom and by his astonishing ability to get things done under difficult conditions. For me, the Pontifical Academy *was* Chagas. A mutual feeling of trust and friendship developed between us. One rarely makes new friends at an advanced age, but my relation to Chagas was just that. He was always sympathetic and understanding even when I was opposed to some of the official tenets of the Catholic church. I have always been especially happy that we were able to work together well during the difficult time when we tried to formulate an effective position of the academy on peace and the prevention of nuclear war. It was Chagas who convinced the pope not only to support our positions but to develop them into an effective promotion of peace and disarmament around the world.

Another Catholic to whom I became closely attached in the course of my activities with the Pontifical Academy is Franz Koenig, the former cardinal of Austria. He took a most courageous stand against the Nazis during the occupation of Austria. He once described to me the attacks of Nazi gangs on the offices of the Catholic church and showed me the bullet holes in an old religious picture in the Episcopal Palace in Vienna. For a time he was the Vatican's advisor in questions of interfaith relations, and he was involved in some of the academy's peace efforts. We had many discussions about these problems. Cardinal Koenig was unusually open to questions about many of the subjects normally considered off-limits by the Catholic church. I was even able to talk to him about the controversial subject of birth control. My contention was that since medical progress has led to "death control," it is our duty to use birth control in order to maintain the natural equilibrium between humankind and nature. In religious terms this can be formulated by saying that God created a natural equilibrium, which science has thrown out of balance. Therefore, it is our duty toward God to reestablish that balanced scheme of things by using birth control. In our conversation the cardinal approved of that argument, but it was certainly not a point of view the Vatican accepted.

After 1985, I became less interested in the activities of the Pontifical Academy. It seemed to me that the pope was no longer as actively interested in the nuclear question. Perhaps he felt that he had already said what he had to say and wanted to avoid taking a stand on detailed political issues such as SDI. I always felt, as did President Chagas, that the academy was supposed to be interested in relations between science and society and therefore must deal with the military applications of science. I had the impression that the pope was worried that the academy had become too politicized. However, the danger to our environment posed by industrial pollution and the destruction of forests will be an important subject for the academy in the near future.

Unfortunately, Chagas grew older, and his health became not very good. This fact, as well as opposition on the part of certain people in the Vatican, combined to force his retirement as president in 1988. Without his influence the academy will not be the effective progressive, liberal force that it has been in the past decade. When Chagas retired, I wrote to him, "God has given us a heavy task. I believe we did accomplish some of it, perhaps not enough, but enough to feel good about it."

In connection with my association with the Vatican, I came in touch also with Cardinal Humberto Medeiros of Boston. He was very interested and active in spreading the knowledge about the danger of nuclear war. When he found out that I was a member of the academy, he invited me to participate in several conversations with him and his colleagues. His ideas were close to those of the *Pastoral Letter of the American Bishops* published in 1983, which emphasized the moral problems of the current nuclear policy of deterrence and of the arms race. Medeiros created an award to honor persons active in fighting the nuclear dangers. The first recipient was Bernard Lawn, the cardiologist, who is a founding member of the organization International Physicians for the Prevention of Nuclear War, a group that received the Nobel Peace Prize in 1986. At Medeiros's invitation I gave the address and presented the medal to Lawn at the award ceremonies.

When Cardinal Medeiros died, he was succeeded by Cardinal Bernard Law, who has shown no great interest in the problem of avoiding nuclear war. When I did meet with him once, together with George Williams, a professor of theology at Harvard, I had a great deal of trouble convincing him that the nuclear problem should be important to the church. Although he understood the danger of a nuclear conflict, he was more alarmed by abortion, since, as he put it, there are millions of souls killed every year in this way.

My connection with the Vatican necessarily brought up the problem of science and religion, and once again I called on the teachings of Niels Bohr and his concept of complementarity: There are seemingly contradictory ways of speaking about human experience that actually illuminate different approaches to the same topic. Religion is a special way of talking about the greatness of this world and about our duties to humankind. Although I would not myself use many of the words and the kind of mythology that religion presents, it is an impressive way of describing the great and the miraculous features of the world in which we live and the ethical standards of human behavior that are the preconditions of a viable society.

The story of the resurrection of Christ is an example. Obviously, it could not have happened in everyday reality, but it is an extremely effective way of emphasizing the validity of his teachings. It elevates the significance of Christ in much the same way that the music of Bach's *St. Matthew Passion* celebrates the words of the scripture. The following anecdote may further illustrate my point. In a Jewish theological seminary, discussions are being held for hours on end to find proofs for the existence of God. Finally, one rabbi gets up and says, "God is so great He does not even need to exist." Existing is a category appropriate for objects dealt with in daily life and in scientific observation, but the categories of the religious approach are in a different realm. What may be sensible and concrete in one frame of reference appears tenuous and vague in a complementary frame.

A more worldly example of the power of religious symbols was revealed to me in a story I heard from Rabbi Alexander Safran, the chief rabbi of Geneva, whom I met in 1963 when he officiated at the bar mitzvah of the son of one of my colleagues at CERN. The incident he described happened when he was the chief rabbi of Romania during the Nazi occupation.

It became known to the Jewish authorities in 1942 that the Transylvanian Jewish population was about to be deported, and there seemed to be no way of forestalling this event. In his desperation, the rabbi decided to try to contact Metropolitan Nicolae Balan, the head of the Romanian Orthodox church of Transylvania, although Balan was known to be an anti-Semite. Balan was in another city, and Rabbi Safran was not permitted to travel. It seemed impossible— in fact, the rabbi calls it a miracle in *Resisting the Storm,* his memoirs— but when the rabbi sent him a message that he wanted to see him urgently in Bucharest, the metropolitan replied that he would come.

At their meeting, Balan seemed cold and unmoved by the rabbi's pleas. As the rabbi talked about the plight of the Jews, Balan remained completely impassive. In his memoirs the rabbi describes how, "impelled by some supernatural power," he found himself yelling at Balan, threatening that he would confront him with the death of thousands of Jews when, on the day of judgment, the two of them stood before the throne of God. This tirade seemed to stir Balan to action, and he made a telephone call that ultimately resulted in the deportation being canceled. As he left, the metropolitan took the rabbi's hand, looked him in the eye, and said, "Remember this day."

Shortly after the war was over, a professor from the Orthodox Theological Seminary of Sibiu appeared at the rabbi's house with a message from Balan, who, the professor said, was in serious difficulty, threatened with arrest and deportation for his wartime activities. "The metropolitan asked me to remind you of a certain day in August 1942," he added. The rabbi immediately went into action. He spoke to the ministers in charge about what Balan had done for the Jews on that day in 1942. Because of his efforts, the metropolitan not only escaped arrest, but he was not tried as a war criminal and was allowed to keep his position as metropolitan of Transylvania.

The German poet Wilhelm Willms has spoken about God in a way that I find very moving; it expresses many of my views about this subject. Here is a quote from one poem, as translated by Douglas Worth:

### God Speaks

I am,
says God,
an impotent god.
Do you believe
I'd have let A-bombs and napalm fall
or people starve to death?
Do you believe I'd have allowed corruption
to flourish wherever you look?
Do you believe
I'd have poisoned the earth? . . .
Do you believe
I'd have built the colossal wheels
of industry
without considering
what might lie in their path?
Do you believe
I'd have divided the world
into two thirds starving
and one third gorged with rich? . . .
You called me omnipotent
but that was to shift the blame
for all that is dark and unaccomplished;
you wanted me to be the devil.
But I'm impotent.
I exist
but only in you
if you want me.
I am goodness

but not without you.
I provide plenty of bread
if you grow wheat
and give enough to all.
My mercy draws tears of relief
if your voice is soft with compassion.
My greatness takes form and life
if your art is inspired.
All things are meaningless
accidents, works of chance
unless your marveling gaze,
as it probes, connects and orders,
makes them divine. . . .
The world is godless
if you are not holy.
How can I be Father, Supporter
unless you honor me
as sons and daughters
by loving and respecting
the God in yourselves?

⊠

# FOURTEEN

# Mozart, Quantum Mechanics, and a Better World

As I CAME TO THE END of this description of my life, I felt a certain regret that my task was over. I had enjoyed reliving a life that has been eventful and interesting and—I cannot deny it—extremely happy. I was lucky to have escaped the Nazi terror, to have chosen a profession with so many international ties, to have had the opportunity of working with the greatest physicists of my time, and to have had their help and support at critical junctures.

My other great good fortune was to have found Ellen, an ideal companion for my life. While I could not share with her most of the joys of insight I received from physics, she understood the significance of these experiences. Her wise and loving care created for me and for our children an environment of blissful serenity. Without her deep understanding of human concerns I do not think I could have fully appreciated Niels Bohr's idea of the complementarity between the scientific approach and other human issues.

Through the power of her loving kindness she was able to transfer to each of us in the family her devotion to everything that is good and beautiful and her convictions that the evils of society can and should be remedied. I consider it an indication that a kind of immortality is possible when I see our children continue to cherish and work toward our ideals in their own creative ways—my son, Tom, as an economist; my daughter, Karen, as an educator specializing in reforming school systems.

\*     \*     \*

Throughout my life, science, the arts and music, and social and political affairs have been my main concerns. The title of this chapter symbolizes these three essential elements of my life. I have already dealt in some detail with science and social issues. But I have not said enough about the role of music and what it has meant to me, so I have found a reason to continue this autobiography for just a few more pages. I will also discuss some ideas having to do with the relationship of science to culture and larger human concerns.

But first music. Here was one aspect of my life that I shared completely with Ellen. Music was a mutual bond and a constant source of inspiration for us. We spent most of our free evenings listening to the records in our large library. From a personal point of view, the development of recorded music is one important way in which technology has contributed to culture. Our lives have been enormously enriched by the opportunity to listen to and study in detail the great creations of music at any desired time. But I must add that I have always been opposed to "background music." How can one converse about trivialities at a party or even engage in serious work if, at the same time, the deepest expressions of life and death, sorrow and elation fill the room?

Throughout my life I have played the piano and participated in chamber music when I have had the opportunity. I regularly played violin–piano duos with Eugene Lehner, whom his friends call "Yano." Because he was a violist with the Boston Symphony Orchestra, Yano considered himself an amateur on the violin. He had been a member of the Kolish Quartet, which was famous for introducing contemporary music to international audiences. Yano came from Hungary and so, like me, had been a subject of Franz-Joseph as a child. Yano helped me understand the deeper significance of music. When we played together, we were restricted to music that was technically simple enough for me, such as Mozart, Beethoven, Schubert, and some Brahms. But he also introduced me to the music of composers I had neglected, such as Berlioz, Verdi, Dvořák, Bartók, and Berg. My parents had had a prejudice against twentieth-century music and, strangely enough, also against Verdi because they considered him not serious enough. How wrong they were! Without

Yano, my relationship to music would have remained much more narrow and superficial. I owe him an enormous debt, as do so many others who were inspired by his loving and enthusiastic teaching, coaching, and interpretation of music. We share another important aspect of our lives: both of our wives are Danish, and Ellen and Lucca Lehner became best friends.

When I speak of the music that filled our lives, I mean the music that began with Bach and Handel, had its culmination in the nineteenth century, and lived on to a certain extent in the twentieth century. It is often called "classical music." (American music such as jazz did not mean much to us, probably because of our European upbringing. Yano expressed it well when he said, "Jazz is not our world, but it has its values, and it is important to recognize that it exists.")

Of course, music as we know it began to develop long before Bach and Handel. It played an essential role in Western culture as part of religious ceremonies and at festive occasions. At the beginning of the eighteenth century, however, the character of music changed radically. It became more individualistic, more expressive of human emotion, and less bound by tradition. Certainly, the transition was not sudden—there are examples of emotional, expressive music in the works of Vivaldi, Monteverdi, and others—but there was a definite change in the compositions of Bach and Handel. In other arts, such as painting and sculpture, this change occurred much earlier. The works of Matthias Grünewald, Albrecht Dürer, Michelangelo, and Leonardo da Vinci are full of individual expressions. I know I will offend many lovers of early music when I maintain that I have not been able to recognize the emotional aspects of this music. Perhaps this is personal prejudice. However, music is enjoyed subjectively, and personal taste is an important part of that enjoyment.

I have often wondered why music was such an essential part of my life. Many mathematicians and physicists are known to be great lovers of music. It has been said that music and the mathematical sciences have much in common, since both employ logical sequences, use applications of symmetry, and depend on the development of an idea from a simple to a sophisticated form. A Bach fugue can be compared to the posing and solving of a mathematical problem. But

I am not satisfied by such formulations. I think that they touch only on superficial common traits. They ignore the fact that art and music on the one hand and science on the other are fundamentally different approaches to human experience. It is just that difference that attracts me to music. I cannot live constantly in the scientific realm. I need the change to other approaches offered by music and the other arts. There is a saying, "In the morning I turn from mystery to reality, in the evening I return from reality to mystery." We need such differing approaches—Niels Bohr's complementarity—as we sometimes need to turn to the other side in bed in order to get comfortable.

Is the music I am considering here really "Western"? Certainly, the composers are mostly from European countries, and their works must be considered as part of Western culture. But there is more to it. We have observed in the last few decades how well this music is understood and performed by Japanese and Chinese artists. To me this is an indication of the universal significance of the music. It expresses human concerns and values that seem to be valid for all cultures. This transnational and transcultural element is expressed in a language accessible to everyone. In other art forms, such as poetry, painting, or sculpture, differences in spoken language, of body and landscape are divisive elements. Music, on the other hand, seems to be a unifying art.

In this respect music is similar to natural science, which also was conceived mainly in the western part of the world but spread to all cultures. Its language, logic, and mathematics are also international. Moreover, it is remarkable that the great expansion of science took place almost at the same time as the expansion of musical style that took place during the last three centuries. Although music and science are basically different cultural phenomena, they are perhaps the most striking examples of two complementary approaches.

At our house, an evening devoted to listening to recorded music started with the difficult decision of what to play. Should it be Bach, with his sophisticated structures in which you discover with each playing some new, intricate facet? Or should we enter a grand musical edifice as constructed by Handel? Perhaps we were in the mood for Haydn or Mozart, who combined great beauty and emotional expressiveness so magnificently, or Beethoven, whose music remains for me the most impressive of all, especially in his later works that touch the deepest and most relevant issues of human concerns. Or, finally, we might play some of the romantic music of the later nineteenth century or some Bartók, Stravinsky, or Alban Berg.

I am delighted by most of Haydn's music with its thematic invention, his zest, joy, and beauty. Still, I agree with Haydn when he said that Mozart surpassed him. Mozart's music contains all of Haydn's wonders, but there is, in addition, an inner logic of consecutive sequences and a tension built up in the development of his ideas that remain unique in music of all times. Nevertheless, I would place Haydn's oratorio *The Creation* among the greatest music ever written. It is full of enthusiasm for the natural phenomena created by God in six days, without any indication of the darker side of existence, except perhaps the warning at the end directed to Adam and Eve: "O happy pair, happy for evermore, if vain delusion lead you not astray to want more than you have and know more than you should!"

In my talks on the origin of the universe, I try to show that the question of the beginning of the world is not only a scientific one. I use scientific language to characterize it as a sudden blast of light of tremendous size and force, the famous big bang. Then at the end of my talk I play a fragment of Haydn's *Creation* to demonstrate another way of describing the same event. We hear a choir of angels singing mysteriously and beautifully, "And God said let there be light." Then, at the words "and there was light," the entire choir and the orchestra explode into a blazing C-major chord. There is no more beautiful and impressive presentation of the beginning of everything.

About Beethoven I can hardly say enough. I cannot count the number of times I have listened to the late piano sonatas and quartets and to the *Missa Solemnis*. Each time, I am excited, distressed, and inspired by the heights and depths of what is expressed in simple, yet extraordinarily sophisticated, ways. A special state of mind is necessary to accept these surges of impressions and expressions of the greatest human emotions. In contrast to some of my friends, I would venture that the late piano sonatas are even more convincing than the quartets. For me, they are the highest musical achievements ever conceived.

In spite of the desperation expressed in so much of Beethoven's music, there is always an allusion to hope and final salvation. His adagios are the nearest thing to genuine prayer. The last part of the *Missa Solemnis*—"Miserere nobis, Dona nobis pacem"—is an expression of hope, albeit tempered by the interspersed threatening military music and by those uncanny tympani beats shortly before the end. In this juxtaposition of salvation and desperation I see a relationship to Dostoevsky. The superhuman dimensions of Beethoven's music also remind me of the impact of Rembrandt's paintings and Michelangelo's sculpture.

Then there is Schubert, whose music also displays that combination of beauty and emotional expressiveness that we find in Mozart, but with a difference. After all, Schubert knew Beethoven's music, and with Schubert the romantic period began. What is so fascinating in Schubert's romanticism is its simplicity and the complete lack of pomposity. Both attributes are the hallmarks of Schubert's songs. I am always astonished and delighted by what Schubert could express with the simplest means. Mozart and Schubert died young. It would be interesting to count the incredible number of notes these masters wrote during their short lives—fantastic achievements both qualitatively and quantitatively.

But I have touched on less than half of our record collection. From Schubert on, the world entered the rich period of romantic music. Chopin, Schumann, and Mendelssohn did not appeal to us that much. We found that they do not quite measure up to Berlioz, Wagner, Verdi, Brahms, Dvořák, and the French impressionists Debussy and Ravel. But some of Schumann's works, such as the great Piano Quintet and his many songs, impressed us also. Brahms is a revelation by himself. The more often I hear his music, the more interesting it seems. He took over much of the heritage of Beethoven but reshaped it in a new personal style that encompasses the whole spectrum of human emotions, from ecstatic exuberance to the subtlest contemplation. Dvořák is more in the Mozart or Schubert line, with his melodic invention, playfulness, and sense of joy.

Altogether, we preferred chamber music, piano, or choral music to symphonies. Even Beethoven's symphonies do not have the same sophistication and depth as most of his other works. A full orchestra may be too robust and often sounds trivial compared to the other forms. This remark does not apply to the use of the orchestra in oratorios or in such masterpieces as Brahms's *Requiem*. And I do not have any doubts about the symphonies of Mozart and Haydn, which I believe equal in every respect the other compositions of those masters.

The French composers of the late nineteenth century exert a special fascination for me. Debussy was one of the true innovators who changed the character of music, just as Beethoven, Bach, Wagner, Stravinsky, and Schoenberg did. By contrast, Mozart was not an innovator. He remained within the style of his time, but his unique genius enabled him to elevate it to levels never attained before—or after. Debussy created a style that is the direct analog of impressionism in painting. Cézanne and the other impressionists invented a new way of looking at nature, in particular at the color of things. Debussy's music places a similar emphasis on the "color" of sound combinations. It is the same soft transition from one color to the other, producing a kindred feeling of pleasure and excitement. Ravel and others expanded his style. These composers opened up new vistas for romantic music.

The last third of our music collection contains the later composers, such as Mahler, Bruckner, Bartók, Schoenberg, Berg, Britten, and Stravinsky. My somewhat reluctant admiration of Mahler and Bruckner does not stem from their almost exclusive use of the symphonic form. I was a great admirer of both composers in my youth, but I have become much more critical. I am no longer greatly impressed by the rather pompous and repetitious Bruckner symphonies, although I find the slow movements captivating in their fervent expression of deep faith. For example, I consider the third movement of Bruckner's Ninth Symphony among the most affecting adagios ever written.

Mahler's music becomes more and more problematic for me. Although I was extremely impressed by all his work for a long time (perhaps with the exception of the Eighth Symphony), I am increasingly disappointed with the harshness and length of many of the movements of his symphonies and with a certain sentimental triviality I now find in too much of his work. But I still consider Mahler's songs, his Fourth Symphony, and *Das Lied von der Erde* among the greatest music ever composed. The Fourth Symphony, written near my beloved Altaussee, is full of almost Mozartian charm and beauty; its slow movement is so tender and sincere, and its finale sheer bliss. Perhaps I should also include the Ninth Symphony with its serene last movement, his farewell to life.

*Das Lied* encompasses many of the deepest aspects of human life and its relation to nature, expressed with spare but convincing means—spare in contrast to most of Mahler's symphonies. The wonderful poems of Li Tai Po and other Chinese poets add to the significance of the thoughts transmitted by this music. When I hear at the end, "Die liebe Erde blüht auf in Lenz und grünt auf's neu! All überall und ewig blauen Licht die Fernen! Ewig . . . Ewig . . . Ewig . . . (The good earth will green and blossom in the Spring, all over and forever . . . forever)" expressed in such convincing musical terms, I cannot repress the thought that it may no longer be so sure in view of pollution and the possibility of nuclear war.

Although Schoenberg is often considered one of the modern composers, his early works actually belong to the romantic period. His oratorio *Gurre Lieder* is a piece we particularly treasured. Schoenberg set to music a cycle of poems by J. P. Jacobsen, a nineteenth-century Danish poet. Perhaps my deep sympathy for everything Danish has played a role in this preference. The forceful poems, which Schoenberg used in a German translation, deal with the ardent love of King Valdemar (a twelfth-century Danish king) for his sweetheart, Tove, and with the bitterness and sorrow of Valdemar after the queen poisons his mistress. The music is more than adequate to the poetry. It is in some ways the last statement of the romantic era that began with Schubert and ended with Mahler.

Schoenberg broke with the romantic tradition and influenced the next period of music. I do not have much empathy with Schoenberg's later works, but I greatly admire his pupil Alban Berg. Berg's Violin Concerto is one of my favorite pieces of modern music, perhaps because he was able to infuse the atonal idiom with a good deal of romantic character. For me, the emotional romantic element is indispensable. That is why I am less attracted than I perhaps should be to Stravinsky's music, although his "Oedipus Rex" and the *Histoire d'un Soldat* impress me greatly. Perhaps I sense some grains of romanticism in these works.

I was introduced to Béla Bartók when a friend got me a ticket for a concert performance of "Bluebeard's Castle." This early work is still somewhat under the influence of the romantic tradition, but the personal style and forcefulness of the music awakened my interest in Bartók's other works. After listening to his quartets, his Cantata Profana, and his piano concertos, I had to admit that he was one of our greatest composers. I was sometimes taken aback by his tendency to excessively harsh rubatos, which prevent me from understanding the intention of some of the pieces. This is why I admire his later works, such as the Third Piano Concerto, which is largely free from harshness.

Contemporary music attracts me because its language is so different from that of the music of the eighteenth and nineteenth centuries. It is almost, but not quite, as opposed to the older music as the scientific world is opposed to the arts. Perhaps a kind of complementarity exists between modern music and that of the earlier periods. I feel that some composers of the modern era are just trying to be different, but in the works I've mentioned and in some of the creations of Benjamin Britten, Frank Martin, Paul Hindemith, Arthur Honneger, and Gyorgi Ligeti, I have found much that opened my mind to new ways of facing human experience in musical terms. My personal friendship with Ligeti definitely helped me enter into this new musical world.

And finally to opera. An opera is a strange creation. One would not have believed that a dramatic theater with singing actors could develop into a serious art form. It seems rather unnatural to have actors sing their roles rather than recite them in prose or at least rhymed verse. The essential point is the great power of information that is delivered by music compared to the spoken word. This fact was brought home to me vividly when I attended a mass at one of the meetings of the Pontifical Academy. I heard the priest speak these familiar words in Latin: "Gloria in excelsis deo, et in terra pax hominibus bonae voluntatis"; "Credo in unum Deus factoren coeli et terrae, visibilium omnium et invisibilium"; and, of course, "Agnus Dei, dona nobis pacem." Immediately the surging chords of Bach and Beethoven came to my mind so intensely that the spoken words seemed dry and empty without the music that accompanies them in the great religious compositions and endows them with much greater significance.

The great power and persuasion that music adds to the spoken word is the basis of the opera as an art form. Indeed, many great operas have worthless librettos that do not diminish the power of an opera by a great composer. Also, most operas deal with elementary emotions such as fervent love, sorrow, hate, and elation. The singing elevates these emotions from the personal to a higher, more universal sphere. The opera setting permits us to witness these often exaggerated, sometimes primitive emotions on an idealized level that transcends personal experience.

I have had a long-lasting exchange of opinion with my friend Isaiah Berlin, who is also a great opera buff and was for some time on the board of the Covent Garden Opera House in London. We have discussed at length the question of "the twelve best operas," and I would like to share our choices—but, with apologies to Isaiah, focusing on my own favorites.

The oldest opera that deserves to be on the list is Christoph Willibald Gluck's *Orpheus and Eurydice*. (I do not include Handel's operas because I consider them rather formal.) The power of Gluck's opera is strikingly evident when Orpheus faces the ghosts of the underworld. They deny him entrance to their domain with several *no's* sung in unison, but Orpheus disarms their resistance with his emphatic, engaging songs. The dramatic return of Eurydice to the underworld and Orpheus's deep sorrow as expressed in Gluck's music are among those great operatic moments in which personal feelings are elevated to a level of universal validity.

Chronologically, Mozart operas belong next on the list. There is no question about *Don Giovanni,* perhaps the greatest opera ever written. There are so many well-known high points in this masterpiece that to list them seems unnecessary. It is hard to find a similarly successful unity of beauty, emotion, tragedy, and drama. *The Mar-*

*riage of Figaro* is second, but then I face the difficult question of which Mozart opera should come third. The limit of three Mozart operas is vital if we are not to crowd out later worthy candidates. The contenders are *The Magic Flute, Cosi Fan Tutte,* and *The Abduction from the Seraglio.* The latter impressed me because of its directness and unity, apart from its other Mozartian qualities. I may have been influenced by my early introduction to its genius by Hans Riehl. In my mind the other two both suffer from their unusually silly librettos. (Isaiah also chose *The Abduction* but added *The Magic Flute.*)

Entering the nineteenth century, I exclude *Fidelio.* There are certain great examples of Beethoven's intense grandeur in this opera, but it does not quite live up to the others on the list. So we come to Verdi. Again one is faced with an embarrassment of riches. The three obvious choices are *Falstaff, Otello,* and *Don Carlo.* Here Isaiah and I agree. It is a miracle that Verdi was able to compose two of his best operas while he was in his seventies: *Otello,* the embodiment of the tragedy of love, and *Falstaff,* the embodiment of sparkling, lusty humor. Reluctantly, we had to leave out *Rigoletto, Un Ballo in Maschera,* and others far superior to so many operas of that period. I am not impressed by most of Puccini, with the possible exception of *Tosca.* I cannot deny that my judgment is somewhat biased toward operas with social or political significance. That is why I include *Otello* and *Don Carlo* with special enthusiasm and regret that *Tosca* has no place on our list. The same reasoning makes me so fond of Verdi's *Simone Bocconegra.*

I admire Richard Wagner in many respects. When I judge his music, I have to disregard his anti-Semitism, which I am convinced had no influence on his work. I am truly fascinated with only a few of his operas, above all *Tristan and Isolde, Die Meistersinger,* and parts of the Ring cycle, mainly *Die Walküre. Tristan* contains by far the most intense and explicit musical rendering of erotic passion and devotion ever composed. There is nothing else in the world literature of music that expresses longing, yearning, and ardent desire with such force. The overture is as close as one can get to a musical description of sexual union. Not everyone would agree with my enthusiasm for that opera. I am told that Einstein, when offered a ticket for a performance of *Tristan* with Lauritz Melchior and Kirsten

Flagstad, said, "No, thank you. They have already died much too often."

I have put *Die Meistersinger* on the list without any hesitation. The dramatic tension, the mixture of caricature and earnestness, and the believably human characters, especially Hans Sachs and Eva, combine to make an opera that is both captivating and convincing.

The only other Wagner opera that should be included is *Die Walküre,* with its wonderful first act full of youthful love and the third in which Wotan parts with Brunhilde, a testimony to deep fatherly love. It enters the list as the representative of the Ring cycle. The other Wagner operas are too harsh, repetitious, and pompous to meet the standards of our list. Isaiah included only *Tristan,* but neither *Die Meistersinger* nor *Die Walküre.* He ascribed my choice of them to my German background.

Wagner was undoubtedly an innovator. He introduced new tonal combinations—think of the introductory motif of *Tristan*—and new uses of orchestral sound, especially in the wind section. These innovations had a decisive influence on Brahms and other composers in the second half of the nineteenth century. Wagner wrote the words to his operas as well as the music, and one has to be familiar with the German language to appreciate them fully. The English translations that are found in the booklets accompanying recordings are usually incredibly bad. The text without the music, even in German, appears bombastic, inflated, and exaggerated. But the music has the power of amplifying and depersonalizing the sentiments to such an extent that the linguistic excesses are no longer disturbing. The tremendous exaltations of love and devotion contained in Isolde's "Liebestod" require words that would sound unacceptable in a spoken sentence. Perhaps it is only in *Die Meistersinger,* which deals with more mundane events, that the text on its own is not only acceptable but is even, indeed, poetically valuable.

This leaves two slots empty in our list of the twelve greatest operas. Isaiah and I agreed that these should be filled by Modest Mussorgsky's *Boris Godunov* and Berg's *Wozzek. Boris* is great from beginning to end, but the scene in the pub when Grigory flees over the frontier, the scenes of the people in the woods greeting his return, and the scenes of Boris's desperation are all outstanding. *Wozzek* is the only twentieth-century opera on our list. In spite of its atonal idiom, it is unusually impressive. It demonstrates Berg's ability to fuse romantic sentiments with atonality. The passages in which Marie complains about her fate and the terrifying spectacle of Wozzek's suicide are examples of these triumphs. I was also influenced by the sociopolitical significance of the opera's libretto. It is astounding that the play by Georg Büchner on which the opera is based was written in the 1830s. The exploitation of a poor individual is explored with the social consciousness of the 1930s, when the opera was composed.

There are other twentieth-century operas, but they do not live up to the level of the others on our list. Richard Strauss is well worth listening to, but his music is often too overwhelmingly opulent. I consider his first opera, *Salome,* the most rewarding of his works. The famous *Rosenkavalier* is very pleasant, especially to a Viennese ear, but I cannot help feeling that it contains too much Viennese sentimentality. Still, many of my friends would consider this judgment too harsh.

Benjamin Britten's operas are certainly remarkable, and I am greatly moved by *Peter Grimes* and *The Turn of the Screw.* Stravinsky's *Rake's Progress* does not seem to me to be as significant as his other works. And there are many other enjoyable contemporary operas that would bear mentioning if there were no limitations on space and time.

The experience and satisfaction of listening to or playing music are very different for me from the experience and pleasure of studying a scientific problem. Certainly, both include a joy of insight, but the insights are into completely different worlds. When I hear the opening of a piece of music, I enter a realm whose values have nothing to do with the scientific world. Probably it will remain impossible to explain with scientific certainty why particular sound combinations are so much more significant than others. Even if science succeeds in quantifying this mystery, it will be irrelevant to our understanding or enjoyment of music.

In this discussion of my relationship to music I have tried—not always successfully—to avoid the word *beautiful*. Beauty has many aspects. It certainly plays an essential role in any art form. The music of Haydn, Mozart, Beethoven, and Schubert contain much that can be characterized as beautiful. But is Beethoven's Grosse Fugue beautiful? Could you call the last movement of the Hammerklavier Sonata beautiful? I would say no, but they are both deeply significant pieces of music.

It is very questionable to equate beauty in music—or in painting or sculpture—with the beauty of a scientific insight. They differ markedly. What beauty means in art is rather obvious and does not need to be described in words, but it is only one measure of a work of art's value. Significance is more important, and need not be coupled with beauty. Significance can come from the expression of feelings, thoughts, and emotions in new and often unexpected ways and with strength and immediate impact, far above our personal thoughts or emotions.

What is usually referred to as "beauty" in science has little to do with beauty as we experience it in art. A typical example is the Maxwell equations, which in a few lines of rather simple mathematical relations contain the essence of practically all electric, magnetic, and optical processes. They are significant because they encompass such a vast realm of phenomena. We physicists call them beautiful, but it is not the beauty, say, of the beginning of the Mozart Symphony in G-minor. Most of the scientific insights that are considered beautiful bring together seemingly unconnected phenomena in a surprisingly compact way or express the fundamental features of a large number of natural occurrences with one single system of thought. John Keats, in "Ode on a Grecian Urn," said,

> Beauty is truth, truth beauty—that is all
> Ye know on earth, and all ye need to know.

On the contrary, I think many kinds of truth and beauty exist.

Significance plays an equally important role, both in art and science, but again, the kind of significance differs greatly. In science it deals with external phenomena in nature. Even in neuroscience when human feelings and emotions are studied, they are observed from "outside" as the effects of neural interactions. A significant scientific insight is an objective experience and must be reproducible, checked by repeated tests. In art, significance is a subjective experience, unique and characteristic of the one masterpiece that expresses it.

In science, the specific way an insight is expressed by the author is of minor interest except to the historian of science. In fact, the significance and beauty of the Maxwell equations can be appreciated much better by reading some of the later presentations rather than the original. Scientific ideas often become clearer with time, and more modern representations bring out the essentials, unencumbered by unnecessary details.

This is not so in art. Only the original or a faithful reproduction (an interpretation in the case of music) can transmit to the beholder the true significance of an artistic creation. In art it is impossible to separate form from content, whereas it is often pedagogically useful to do so in science. The way the content is expressed plays an essential role in art; indeed, it is what separates art from mere description. Any change in the manner of expression would alter and weaken the content of a work of art. I remember a literature teacher in high school who asked us to repeat a poem by Goethe "in our own words." What a ridiculous request!

In contrast to works of art, scientific creations do not stand alone and cannot be regarded as separable entities. They become part of a single edifice that is collectively assembled by scientists, its significance and power based on the totality of the contributions from many sources. This is referred to as the scientific worldview. Sir Isaac Newton once said in a letter that he gained his insights by "standing upon the shoulders of Giants." His work, like the work of Einstein and other great scientists, comprises only a few stones of this edifice—albeit rather large ones, at pivotal locations.

It must be said, however, that something like a collective edifice of achievements also exists in the arts. Tradition develops from one period to the next. Mozart could not have written his music without the achievements of Haydn and the development of baroque music since Bach and Handel. Schubert and Brahms would not have created their masterpieces without Beethoven. Michelangelo's art builds on Greek art and that of the early Renaissance. We understand a work of art much better when it is considered within the cultural framework of its time. Art grows from a cultural soil fertilized by previous creations. In this sense the artist also stands on "the shoulders of Giants."

The scientific culture differs from the artistic in another respect. There exists something that may be called "scientific progress." We definitely know and understand more today than we did before. Einstein's theory of gravity is nearer to the "truth" than Newton's. If Newton were alive today, he would freely—and probably enthusiastically—admit that Einstein's theory is an advance over his own (a statement that is hard to prove yet is convincing to any scientist). No comparable progress can be found in art. There is no reason to consider a Gothic sculpture better than a Romanesque one, to think Raphael represents an advancement compared to early medieval art, or to see Mahler as more valid artistically than Mozart. It is true that there is a tendency toward increased sophistication in art over time. In general, the means of expression become more manifold, varied, and intricate. Of course, a similar tendency exists in the sciences, but in science it is connected to a genuine increase of insight into nature. The increased sophistication of art may lead to a wider scope of subject matter and a greater variety of creative forms, but hardly to more powerful forces of artistic expression.

Let us look back at the developments and changes in science, in art, and in the whole social structure during my lifetime. The new insights won by science into the workings of nature have been extremely far-reaching and profound. In physics and chemistry the new era started with the development of quantum mechanics and the theory of relativity, which provided the basic ideas for an understanding of the properties and the behavior of matter as we find it on earth and in the stars. What a difference between today's insights into material structure compared to those when I was a student! At that time the application of quantum mechanics to the properties of matter had just begun; today it has developed into so many specialized fields that a single scientist can be acquainted with only a small part of the vast body of knowledge.

The progress of biology provided us with an understanding of the most important biochemical processes and of the molecular mechanism of heredity. Geology and astronomy encompass much wider fields of knowledge and much deeper insights because of the many new methods of observation that contemporary science has provided. Today we look at the sky with new instruments that enable us to see infrared, ultraviolet, and X-ray radiation, as well as neutrino emissions. In geology the development of new types of magnetic analysis, laser techniques, and much more sensitive seismographs has led to a more detailed knowledge of the inner structure and history of our planet.

These numerous and impressive insights are the crowning achievements of developments in the last four or five centuries. Previously, the urge for an understanding of the world in which we live was met with general mythological, religious, and philosophical ideas that delivered "holistic" answers to such fundamental questions as, What was the beginning of the world? or What is life? The answers were directed at the totality of phenomena, attempting to account for everything.

Then, a few centuries ago, human curiosity took a different turn. Instead of getting at the whole truth, separate phenomena were studied. Instead of asking, What *is* matter? What *is* life? people asked, What are the properties of matter? or How does blood flow in the body? They did not ask how the world was created but how the planets move. General questions were shunned and replaced by limited ones. However, detailed answers created a framework for the understanding of more general questions. Only a renunciation of immediate contact with the "one and absolute truth," only endless detours through the diversity of experience allowed the methods of science to become more penetrating and their insights to become more fundamental. Today science begins to deliver at least tentative answers to the general questions, such as What is life? or What is the origin of the universe? Something like a comprehensive worldview has arisen during my lifetime out of a synthesis of the insights gained over the previous centuries.

This scientific worldview differs in two respects from the religious, mythological, and philosophical worldviews of the past. First, it does not contain concepts connected with the human soul, such as faith, morality, dignity, degradation, happiness, and sorrow, concepts that are all related to religion. It does deal with some of these concepts indirectly, as manifestations of neurophysiological processes in the brain. Recent progress in neuroscience and in biochemistry of the nerve cells promises much deeper scientific insights into the mechanisms of all kinds of feelings and emotions. We begin to acquire means of evoking and influencing these psychological phenomena. I am convinced, however, that important aspects of the human psyche will never be touched by such inquiries. They distinguish themselves from most scientifically treatable phenomena because they are unique and nonrecurring. In general, they are what we consider most relevant to our lives.

Second, the insights of the scientific worldview are tentative. They are considered incomplete perceptions, parts of a greater truth hidden in the plenitude of phenomena. The insights are not based on dogmatic principles, revealed to us by divine inspiration or by some internal sparks of full recognition. What is perceived as scientific truth is steadily revealed in partial steps, sometimes big ones, sometimes small ones, and sometimes even steps backward. Some recent knowledge will turn out to be mistaken. Although only a few past insights have turned out to be outright wrong, some others have proven too limited, not general enough, misconceived, or awkwardly formulated. Still others will no doubt appear irrelevant in view of future insights.

The scientific worldview has strongly influenced the general public's attitudes and thoughts. Most people are now aware that natural phenomena follow general laws. This awareness helped diminish the influence of dangerous superstitions, but it did not entirely do away with them. There are still far more astrological than astronomical books addressed to the general reader.

The strongest effect of science on human society comes from the applications of scientific insights to technology. Such applications had led to a pervasive social regrouping through the growth of industry and commerce. With the emergence of the working class a major portion of the burden of social responsibility has shifted from the aristocracy to industrial leaders. The world of agriculture has also been thoroughly changed. In previous centuries, 80 percent of the world's population worked on the land. Now, because of mechanization and the so-called green revolution in agriculture, in the developed countries only 5 percent or less of the population are engaged in agricultural pursuits.

These developments have had both favorable and unfavorable consequences. It has become fashionable to forget the beneficial side and to emphasize the evils of progress. Machines have made strenuous physical labor almost unnecessary. Traffic and transportation have been completely transformed. Medicine, the technology of biological science, made it possible to eradicate epidemics and to double the human life span. Furthermore, because of technological advances, it is theoretically possible to abolish poverty and to feed, house, and clothe the world's population adequately. Of course, these possibilities can only be partially realized even in the developed countries, where the homeless and impoverished number in the millions.

What about the unfavorable consequences of the applications of science? Often they stem from a lack of foresight and from a thoughtless proliferation of technology without consideration of the consequences. The pollution of the atmosphere is a case in point. We have discovered—almost too late—that while industrial development has enriched the developed countries, it has harmed nature. This has happened in spite of numerous warnings by experts about the effect of the by-products of production and the burning of fossil fuels on our globe's atmosphere. Only recently have there been some efforts to correct these problems and to prevent them in the future.

If advances in medical technology that prolong life are not accompanied by corresponding birth control, the resulting population increase will make it impossible to fulfill the promises of technological achievements. Social and religious prejudices continue to impede this obvious measure in just those areas of the world where it seems most necessary.

The worst abuse of scientific knowledge is the ongoing development of increasingly deadly weapons such as nuclear missiles and new forms of chemical-biological warfare. The nuclear arms race has led to the deployment of over 50,000 nuclear warheads ready to be launched. Never before has the world been so threatened with the possibility of ultimate catastrophe. Fortunately, just as this book is in its final stages, a new spirit seems to be emerging. The world's leaders and the general populace are beginning to realize the madness of accumulating these means of mutual destruction, and there is hope of developing a more rational attitude toward our problems. Perhaps a time is coming when the nuclear arms race of the past decades will be regarded as a serious case of collective mental disease that was cured just in time. If this happens, I will be able to have a clearer conscience about my participation in developing the atomic bomb. Our original hope that the terrible destructive power of the bomb would make future world wars impossible may not have been in vain. I am grateful that I have been allowed to live long enough to witness the beginning of these hopeful changes.

Another negative aspect of the scientific/technological worldview is what I call "spiritual pollution." This manifests itself as a general devaluation of other than scientific-rational ways of dealing with human experience. The Swiss physicist and philosopher Markus Fierz aptly describes the situation I am referring to: "The scientific insights of our age shed such glaring light on certain aspects of human experience that they leave the rest in even greater darkness." If scientific/technological progress has freed a good part of the population of the developed world from the burdens of hunger, poverty, and oppressive manual labor, we must ask, To what end has this advancement taken place? What does one do with one's life when there is no longer a need to fight for existence twelve hours a day? Individuals are thrown back upon themselves and must find an aim for their lives. Work in the industrial complex is too often mechanical and secondary, providing no sense of personal achievement for the worker. Most employees have little influence on the direction of the enterprise of which they are a part. What becomes of human dignity? Where is the individual's satisfaction in accomplishment and purpose? For that matter, where is the feeling that life has some sense and meaning?

It is hard to say what *sense* means. I once answered the question of whether there is any sense in the universe with, "In the sense that you sense a sense." Another answer is Bohr's paradoxical remark (he loved paradox), "It makes no sense to say the universe has no sense." Some people maintain that scientific insight has eliminated the need for meaning. I do not agree. The scientific worldview established the notion that there is sense and purpose in the development of the universe when it recognized the evolution from the primal explosion to matter, life, and humanity. In humans, nature begins to recognize itself. The awareness of that recognition carries with it a responsibility to maintain and not destroy what has been created. The German philosopher Martin Heidegger said, "Humanity is the shepherd of the world." The same idea has been eloquently expressed by Thomas Mann, as part of *Lob der Vergänglichkeit* (In Praise of Impermanence), written in February 1951 for the radio program "This I Believe":

> Astronomy—a great science—teaches us to consider the earth as a companion of a most insignificant star in the giant cosmic turmoil, roving about at the periphery of our galaxy. This is, no doubt, scientifically correct. But I doubt that such correctness reveals the whole truth. In the depth of my soul I believe—and consider this belief to be natural to any human soul—that this earth has a central significance in the universe. In the depth of my soul I entertain the presumption that the act of creation which called forth the inorganic world from nothingness and the procreation of life from the inorganic world, was aimed at humanity; a great experiment was initiated whose failure by human irresponsibility would mean the failure of the act of creation itself, its very refutation. Maybe it is so, maybe it is not. It would be good if humanity behaved as if it were so.

The lack of awareness of sense and purpose has led culture to become increasingly shallow. When the most important needs have been provided for, the content of life may amount to no more than a desire for entertainment or pleasure. In the extreme, the lack of a sense of the meaning of life may lead to such excesses as the use of drugs. The damage to our society by drugs, with all their terrible consequences, is more threatening today than the receding danger of nuclear war. Can a better system of education help? Can improved schools provide young people with the conviction that life has meaning? It certainly is worth trying.

What is missing in too many individuals is a feeling of deep commitment to a great cause beyond our own personal interest—a cause whose value is never questioned. We certainly find such commitment among a minority of the population who have devoted their lives to the improvement of our society. We find it among the medical practitioners and researchers who strive to diminish sickness and pain. We find it also among those who are working to develop means of making life easier. And, of course, artists and writers seem to have a clearer sense of the meaning of life or are at least searching for it in their creations.

Finally, we find such a conviction among scientists when they are inspired by the greatness of nature and follow their urge to know more about the world. We also find it when they use their scientific expertise for fighting such global threats as the pollution of the environment and the dangers of modern weapons. In these instances the moral/ethical and the scientific/technical approaches to world problems are combined.

Unfortunately, we also encounter deep commitments all too often in religious fanatics who exclude all approaches different from their own beliefs. Religious fanaticism has become increasingly dangerous. It may become as serious to the world as the arms race, since the followers may use modern weaponry to increase its influence.

By and large, nonscientists are not inspired by scientific insights. What they understand of science nowadays is roughly this: Because everything follows the laws of nature, we do not need a God. Science is appreciated mainly because of its technical applications. In part, this is the scientists' fault. They have achieved great things, but they have not made sufficient efforts to convey the greatness and wonder of these ideas to others in a comprehensible way. I believe that conveying these insights is possible, but it is a difficult task that far too few gifted people tackle. The situation is equally bad in respect to the great achievements of technology. It is true that the proliferation of science fiction helps in some ways to excite enthusiasm for the wonders of technology. (*Science fiction* is a misnomer; most examples of it should be called "technology fiction.") Still, the most prevalent reaction to technological advances has been accommodation to each change that makes life more comfortable, coupled with the fear that further innovations will lead to more deadly weapons and to the destruction of the environment.

Can art and literature provide a sense of purpose to our lives? When art was still in the service of religion, it was no doubt generally comprehensible and acknowledged by the majority. When religion lost its influence, art acquired independence even as it continued to be the expression of the great ideas of each time period. But it was accessible only to a restricted stratum of society. Where are the art and literature that concern themselves with the ideas of science and technology? Surely it is the task of artists and writers to bring the great ideas of our time to the public. I believe that Sinclair Lewis's *Arrowsmith* was the last great novel to describe the excitement of scientific research. Although contemporary art and literature are full of new ideas and impressive examples of original creativity, they do not get close enough to the positive ideas of our culture and do not satisfy our hunger for sense and purpose in the universe.

During my lifetime the spirit and character of science have changed in some respects; in other ways these have remained constant. I will confine myself to examining the changes in physics, because I know it best and because it is basic to all the other natural sciences. The fundamental urge that drives physicists to indulge in finding out how nature works has not changed. This drive is what still causes people to choose that profession from among the many other professions that promise more material advantages with far less effort.

But physics has become a much wider field in the sixty or so years that I have been involved in it. In the early years it was possible to become acquainted with most of the frontier problems. Atomic physics, solid and liquid state physics, and physical chemistry were not separate fields in those days. We were aware of the important experiments and theories in all of these areas of investigation. Nuclear physics did not exist until the 1930s. There were areas of physics such as hydrodynamics and acoustics that, to some extent, lay outside the direct interest of the group to which I belonged. (This did not stop Heisenberg from writing a paper on turbulent liquid flow, however.) The total number of physicists was very small compared to the present.

The most important recent change is a growing trend toward specialization. Today there are so many physicists and so much accumulated knowledge that one could not possibly be well acquainted with most frontier achievements. Furthermore, some branches of physics, but by no means all, rely on heavy and complicated machinery and instrumentation, including the extensive use of computers. Most experiments can only be performed by collaboration between many people working in large teams, especially in particle physics, space physics, and to a growing extent, in nuclear physics and astronomy.

Today a single publication may have more than 300 authors. It would seem that the personal involvement in such work would suffer, since each team member contributes only a narrowly ascribed task. Usually, one or only a few members of such teams are the true leaders and organizers of the work. This requires a certain type of person who not only must be an excellent physicist with a good understanding of engineering but must also be adept at raising the money for these very expensive endeavors and be skilled at keeping a large, sometimes international group working together even if they have differing ideas of how to pursue the project. Consequently, a new breed of physicists has emerged with personalities that are considerably different from—and sometimes not as attractive as—those of the leading figures I met in my earlier days. However, in atomic, molecular, and condensed matter physics most of the experiments are still executed by relatively small teams consisting of several professors working with a few graduate students. These physicists are lucky, but, of course, their focus is not the innermost structure of all matter.

The trend to big teams has not touched theoretical physics. Collaboration between a few people, usually not more than three or four, happens more frequently because of the increasing mathematical sophistication. There is a danger of being led astray by complicated mathematics if one does not have close collaborators who can check each step. Usually there is no obvious leader in such a theoretical team, except when the collaboration involves one professor with his or her students. The increasing use of computers for the evaluation of complex mathematical formulae also requires collaboration among several people. Frequently it is necessary to use computers for the evaluation of theories that are supposed to explain certain phenomena. It is often the only way to find out whether a theory is right, but a reliance on computers alone carries dangers.

Here is an example. I was and still am puzzled by the fact that the wind blows preferably from the west in our latitudes. Once I asked a famous meteorologist for an explanation. He asked me to come to his office, where he showed me his computer outputs calculating the wind directions, taking into account the solar radiation, the rotation of the earth, and other important facts. "You see," he said, "all the arrows in the middle latitudes point west to east." I replied, "Now the computer understands it, but what about you and me?" This is not to criticize the use of computers in theoretical physics; it is often the only way to find out what results from a theory. But we should not be content with computer data. It is important to find more direct insights into what a theory says, even if such insights are insufficient to yield the numerical results obtained by the computer.

Another, perhaps more fundamental, change took place in physics during my lifetime. At the beginning of the century, only a few decades before I became a physicist, there was a lack of understanding of the most obvious properties of objects surrounding us here on earth. Before 1900 one did not even understand why a piece of material becomes first red, then yellow, and then white when it is heated. It was known that everything is made of atoms, but atomic structure was unknown. Furthermore, no one knew why atoms form molecules and solids, why solids are hard to deform, or why metals conduct electricity and heat. The development of quantum mechanics gave us the key to all those questions. In the first half of the century most of the principles needed for an understanding of material behavior under normal conditions were found. But even today much work is still needed before we have a thorough understanding of the details.

When the immediate questions were answered, a subtle change took place. The interest of science shifted toward the properties of matter under unusual conditions, such as very low or very high temperatures, extreme pressure, or extreme dilutions. When new kinds of materials such as plastics or complicated chemical compounds were invented, their properties were studied and, if possible, explained. Some of these studies were not only of great intrinsic interest but also of practical use for technology, as in the superconductivity of metals at low temperature or the properties of plastic materials. Some fields of physics became increasingly involved in technical applications; a good deal of what was originally basic physics became applied physics.

Nuclear physics, for example, deals with matter under unusual conditions. Nuclear reactions do not take place on the surface of the earth except at the targets of manmade accelerators. In nature, nuclear phenomena occur in the center of stars and in cosmic cataclysms such as star explosions. Indeed, the natural radioactive substances found on earth were produced billions of years ago when terrestrial material was ejected by a supernova explosion. They are, as it were, the embers of the fire in which our elements were created. The situation is even more extreme in modern particle physics. In order to penetrate into the innermost structure of matter, substances must be exposed to extremely high energies that, in nature, would have occurred only at the beginning of the universe. Physics has thus performed a leap into the cosmos.

These changes do not detract from our interest. On the contrary, the behavior of matter under extremely high energies has told us much about the fundamental constituents of matter and about the forces between them that cause matter to behave as we know it and will tell us more about this in the future. Lately, particle physics and cosmology have become related, since the temperatures at the beginning of the universe were so high that matter was decomposed into those fundamental constituents that were discovered in particle physics.

Of course, the behavior of matter under any kind of unusual condition is of intrinsic interest. Very frequently the special instrumentations required for such research provide fascinating ways of gaining better insights into problems of other fields of science, and can be important when dealing with the vital issues of modern civilization, such as methods for preventing the pollution of the atmosphere.

One recent field of investigation that is exciting in its own way is the study of chaotic behavior. In previous periods physicists concentrated on regular, repeatable phenomena in order to get at the fundamental laws that govern nature. But complex situations do exist that are not repeatable because very small changes in the external conditions produce large differences in behavior. Such situations are what one calls "chaotic." Yet under such circumstances some regularities have been found that continue to defy explanation. There is certainly much work to be accomplished in this field.

Altogether, physics does not seem to have become less interesting since I embarked on my career. The changes in the character of research are the effects of the successes of our science. The leap into the cosmos and the concentration on unusual conditions are consequences of the fact that regular terrestrial phenomena under ordinary conditions are already understood to a large extent. But physics has lost some of its immediate relevance to direct human experience. With a bit of cynical exaggeration one could say that the realization of those unusual conditions costs us as much as the efforts to solve the problems they pose.

The instruments by which we investigate nature have become more and more sophisticated and complicated. The intellectual distance between the observer and the observed objects has greatly increased, and some of the immediate relationships between scientist and nature have been lost.

Several years ago I received an invitation to give a series of lectures at the University of Arizona in Tucson. I was delighted to accept because it would give me a chance to visit the Kitts Peak astronomical observatory, which had a very powerful telescope I had always wanted to look through. I asked my hosts to arrange an evening to visit the observatory so I could look directly at some interesting objects through the telescope. But I was told this would be impossible because the telescope was constantly in use for photography and other research activities. There was no time for simply looking at objects. In that case, I replied, I would not be able to come to deliver my talks. Within days I was informed that everything had been arranged according to my wishes. We drove up the mountain on a wonderfully clear night. The stars and the Milky Way glistened intensely and seemed almost close enough to touch. I entered the cupola and told the technicians who ran the computer-activated telescope that I wanted to see Saturn and a number of the galaxies. It was a great pleasure to observe with my own eyes and with the utmost clarity all the details I had only seen on photographs before.

As I looked at all that, I realized that the room had begun to fill with people, and one by one they too peeked into the telescope. I was told that these were astronomers attached to the observatory, but they had never before had the opportunity of looking directly at the objects of their investigations. I can only hope that this encounter made them realize the importance of such direct contacts.

A few years ago, my grandson Marc was in his second year at Brown University, intending to concentrate on physics and astronomy. He asked me whether I would choose physics again if I were his age. I thought about it and had to answer no. What drove me into physics more than sixty years ago was the relevance of the unsolved problems that existed at that time to immediate human experience. Today I probably would want to study how the brain functions, what underlies memory and consciousness. We know as little about that now as we did in my youth about the properties of our immediate environment. A year later Marc told me that he had switched to neuroscience. I was very glad to hear it, but I have had an uneasy feeling of having influenced him too much.

The scientific approach to human experiences must be considered as one specific way to deal with them. There are others. When we admire a sunset, we may think of the scientific reasons why the sun is red and the surrounding sky assumes the complementary colors. But we may also be impressed by the beautiful color combinations, or we may consider it a symbol of the end of a day's work. Likewise, the starry sky on a beautiful night may make us think about the fascinating facts of stellar evolution, but it is also symbolic of awe-inspiring infinity and the mysteries of the night. A Beethoven sonata may be considered a combination of sonic vibrations or of certain sensations received by neural synapses, but it also contains an immediate musical and emotional message.

All of these examples grow out of how Niels Bohr extended his concept of complementarity to human experience outside of physics. It was from Bohr that I learned to recognize the essential role of different complementary ways of dealing with the world around me. The following true anecdote is a good illustration: Heisenberg and Bloch were walking along a beach on a sunny day. Bloch was explaining a new way of looking at the geometrical structure of space, when Heisenberg, his mind drifting into other complementary avenues of experience, interrupted: "Space is blue and birds are flying in it."

John Keats, in his poem "Cold Philosphy," wrote:

> Do not all charms fly at the touch of cold philosophy
> There was an awful rainbow once in heaven
> We know her woof, her texture
> She is given in the dull catalog of common things.

I disagree. For me, all the different aspects of an event add to its significance. I consider it more attractive, more interesting, more awe-inspiring, and more beautiful in all the meanings of this term when scientific insight is included in the contemplation of the phenomenon.

Each complementary approach has a specific kind of discourse: it appears lucid and concise within its own intrinsic scale of values but fragile and indefinite when judged by the peculiar requirements of a complementary approach. Poetic expressions would make little sense in a scientific description; compassion has no place in justice; psychological arguments make little sense from the neurophysiological point of view. Still, one aspect complements the other. We must use them all to understand the full significance of our experience.

Unfortunately, most people resist the complementary view of things. There is a trend toward clear-cut, universally valid answers, excluding different approaches. For example, today the scientific approach is often considered the only reasonable and serious one. On the face of it, there are good reasons for this belief. No field of human experiences seems to be inaccessible, in principle, to scientific study and comprehension, although much still is not understood. Science may have a justified claim to completeness. But it must be emphasized that *complete* is not synonymous with *all-encompassing*. Even if we reach scientific comprehension of the processes underlying thought and emotions, it will still be necessary to use other methods of discourse to deal with our complex and varied experiences. A system of thought can be complete within its framework but still omit important aspects of complementary character. Sometimes these aspects are the most relevant ones. Some of the prejudices against science and technology come from a half-conscious resistance to the implicit claim of completeness.

History has shown that, whenever one approach is developed in great force, others are unduly neglected. In the Middle Ages the religious approach dominated. Today science and technology are in the forefront. In medieval times religion's way of looking at the world was so dominant in Europe that nobody noticed the appearance on 4 July 1054 of a supernova that was ten times brighter than the brightest star. There were many monks who registered all kinds of events. However, the prevailing mood was such that no importance was given to a new star. But in China at that time a detailed record of the phenomenon was made, and the decrease of the intensity of its light was recorded as time went on. The scientific approach will prevail only if the civilization considers it relevant.

The Middle Ages' one-sided emphasis on religion and our obsession with scientific/technological developments have both released enormously powerful creative impulses. In medieval times these were focused on art, architecture, and moral philosophy. The Gothic cathedral was the greatest example of the creativity of that period. But the one-sided emphasis led to serious abuses, such as the murderous Crusades, the neglect of corporal suffering, and the low value put on individual lives.

When my students ascribed the cruelties committed during the Vietnam War to the spirit engendered by the scientific/technological ideology of our day, I told them about the conquest of the city of Béziers by the French armies in 1205. The general asked the papal delegate, Abbot Arnoud de Citeaux, what he should do with the population. The abbot answered, "Kill them all; God will select the ones that will go to heaven and those that are damned to hell." Of course, the students pointed out quite correctly that there can be abuses as well in an exclusive emphasis on technology and science. It has led to our thoughtless exploitation of nature, an overemphasis on material values, the "greeding" of the Western world, and the irrational overproduction of weapons of destruction.

It is clear to me that the decisions we need to make and the tasks we face call for an approach that seeks varied and complementary solutions. In order to deal with the totality of human experience we need more than the answers provided by science. We must apply those concepts that deal with the human soul and that apply moral values to the issues we are addressing. For moral and political decisions, scientific insight can be most useful for pointing out the consequences of certain actions, but the decisions themselves always rest on nonscientific arguments.

The call for varied approaches is by no means relativism in the sense that everything is justified. Neither is it a denial of values. On the contrary, ethical principles and a value system must be derived from many sources in order to include openness, tolerance, and a full understanding of the situation.

Human experience encompasses much more than any one given system of thought is able to express within its own framework of concepts. We must be receptive to the varied and apparently contradictory ways of the mind when we are faced with the realities of nature, our imagination, and human relations. There are many modes of thinking and feeling, and each of them contains a part of what we may consider the truth. Science and technology comprise some of the most powerful tools for deeper insight and for solving the problems we face. But science and technology provide only one path toward reality; others are equally needed for us to comprehend the full significance of our existence. Indeed, those other avenues are necessary for the prevention of thoughtless and inhuman abuses of the results of science. The survival of our civilization is severely threatened by nationalism, religious fundamentalism, and other intolerant, one-sided views. We will need to use many different approaches in order to overcome the grave problems that we face today. Only then do we stand a chance of achieving a better world.

# Index